Acclaim for T...

Isabel Losada writes wittily, ... ways in which we can help ... of us suffering from eco-anx... ... a friend to guide us through the maze of conflicting advice, steer us round the pitfalls and gently support our stumbling steps. Her own stories of frustration and confusion, meltdowns and bafflement, will chime with anyone who has ever tried to do their bit for conservation in a world that seems determined to thwart them at every turn. Told with humour and candour, *The Joyful Environmentalist* is a manifesto of brilliant advice offered with humility and good grace – it is a practical guide to empower us all.

Isabella Tree, author of *Wilding*

There is much to read that discourages both individual and collective action to live with love for each other, the world and ourselves. Much that ever so cleverly separates us and undermines our self-worth as citizens together with the power to change. Isabel encourages both individual and collective action with her honest, humorous account of all the choices we each can make right now today to change the world by changing ourselves. She gave my spirit a lift and my feet somewhere to stand.

Sir Mark Rylance

This is the joy we need in our lives, cutting through the despair to help us find our way in the world.

George Monbiot

'Isabel Losada is a Climate Hero'

The Guardian

Perhaps the most important message any environmentalist can give at the moment is that green options are neither just a last desperate hope for saving the planet, nor misery maximisers that will make life less worth living. They are ways of affirming that worthwhileness and enhancing it. This book, practical and realistic as well as visionary, will keep that positive message before the reader's eyes. Joy is after all one of the best motivations we can have for change.

Dr Rowan Williams

Critical Acclaim for Isabel Losada's previous books

The Battersea Park Road to Enlightenment

'A total delight. Isabel Losada navigates her way through the eccentric highways and byways of the human potential movement with scepticism, humour and interrogative open-mindedness. Candid, thought provoking, sassy and very very funny.'

Mick Brown – *Daily Telegraph*

'Full of a crazy joy ... made me laugh out loud.'

Impact Cultural Magazine

'Swift, Snappy and Engaging.'

The Sunday Tribune

For Tibet, With Love

'The world must be changed ... Isabel's story brings this truism to life, in a vivid, funny, heart-warming, delightful way. It is a great read, a live teaching! I enjoyed it, laughed and learned a lot!'

Professor Robert Thurman, Tibetologist and Buddhist Scholar, Columbia University, New York

'Isabel Losada is a 21st Century Hero ... someone who is changing the world for the better and will make you want to, too.'

Harpers and Queen Magazine

'This remarkable tale of one woman's dedicated personal journey captures the spirit of compassion in action.'

Lama Suyra Das

'Fast, funny and inspiring too. Isabel Losada is a writer that changes lives.'

Joanna Lumley
Men!

'Isabel Losada has achieved the perfect combination of humour, poignancy and intellectual rigour.'

The Statesman

'Think Michael Moore meets *Men are from Mars* – only a lot funnier.'
Glamour Magazine

'Finally, everything we couldn't work out is explained. Brilliant, honest research and funny too.'

Elle

The Battersea Park Road to Paradise

'Isabel Losada proves herself a fantastic prose-stylist and the most eloquent of guides. It's a book which grapples with the big questions.'
**Piers Moore Ede,
author of *All Kinds of Magic***

'Engaging, lyrical and courageous. Her journey and her honesty will make you laugh and touch your heart. This is rock'n'roll of the soul. I love this book.'

William Bloom

'The intriguing thing about this thought-provoking book is the amount of genuine happiness on display ... happiness based on being at peace with yourself.'

The Daily Telegraph

Sensation

'Sensation explores the gentle beauty of intimacy and pleasure in human relationships; it's impossible not to be entertained, educated, amused and inspired.'

The Amorist

'Let's talk about important, deeply nourishing sex that we could ALL be having if we brought more honesty to the bedside table.

This kind of sex conversation is the point of Isabel Losada's new book, *Sensation: Adventures in Sex, Love and Laughter*. Losada leaves no stone unturned – or should I say unturned-on? Nakedly frank and frankly nakedly necessary.'

Caroline Sanderson,
The Bookseller

Thank you

THE JOYFUL ENVIRONMENTALIST

With Love,

Isabel M Losada.

WATKINS

The Joyful Environmentalist
Revised and Updated 2nd Edition
Isabel Losada

First published in 2020.
This edition first published in the UK and USA in 2025 by Watkins,
an imprint of Watkins Media Limited
Unit 11, Shepperton House
89-93 Shepperton Road
London
N1 3DF

enquiries@watkinspublishing.com

Design and typography copyright © Watkins Media Limited, 2020, 2025

Text copyright © Isabel Losada 2020, 2025

Isabel Losada has asserted her right under the Copyright, Designs and Patents Act 1988 to be identified as the author of this work.

All rights reserved. No part of this book may be reproduced or utilized in any form or by any means, electronic or mechanical, without prior permission in writing from the Publishers.

1 3 5 7 9 10 8 6 4 2

Typeset by JCS Publishing Ltd.

Printed and bound by CPI Group (UK) Ltd, Croydon, CR0 4YY

A CIP record for this book is available from the British Library

ISBN: 978-1-78678-979-2
ISBN: 978-1-78678-980-8

www.watkinspublishing.com

For Jack

'Whatever you do will be insignificant, but it is very important that you do it.'

Mahatma Gandhi: 1869–1948

'You have given me a thing I could never have imagined, before I knew you. It's like I had the word "book", and you put one in my hands. I had the world "game", and you taught me how to play. I had the word "life", and then you came along and said, "Oh! You mean this."'

Richard Power, *The Overstory*

Contents

Introduction	1
Plastic 1: Dangerous Woman in Whole Foods without Fork	5
Plastic 2: Ten New Plastic Products Sold to Us on Social Media This Year	12
Earth 1: Plant Your Trees	14
Plastic 3: The Basics: How to Remove Plastic from Your Home – 36 Steps for Starters	38
Technology 1: Our Digitial Footprint: The Illusion of a Cloud That I Recall	50
Energy 1: Smart Meters, Candlelight and Sleeping Children	58
Energy 2: Changing Light Bulbs on the *Titanic*	64
Plastic 4: Soap	68
Food 1: Holy Cows and Fish Fingers	71
Food 2: How to Buy a Tomato – and Other Supply Chain Dilemmas	82
Travel 1: Cars, Bikes and Legs	103
Travel 2: Flying Hypocrisy with Plastic Too	110
Stuff 1: The Life-Changing Magic of Not Shopping	117
Stuff 2: Ethical Consumption: To Buy or Not to Buy	122
Energy 3: Burning Coal and Admissions of Stupidity	133
Energy 4: Gas and Hot Air	137
Energy 5: Smelly Emissions and Profit Margins	143
Wine 1: Cosmic Grapes	149
Animals 1: Myths and Palaeontology	152

Climate Change 1: Flat Earthers and Climate Change Deniers	156
Insects 1: Bug Houses and English Bluebells	163
Earth 2: Alchemy and Black Gold	169
The Bigger Picture 1: Being a Little Bit Activist	173
The Bigger Picture 2: Are You an 'Arrestable'?	213
Birds 1: Vegan Fat Cakes and the Merits of Pyracantha	225
Money 1: Putting your Money Where Your Values Are	234
Earth 3: Befriending Your Urban Trees	239
Energy 6: Gas, Electricity and Solar Panels	255
The Bigger Picture 3: Voting, Canvassing and Wolf Tattoos	277
Earth 4: Yippies and Wombles: Living Off Grid	292
Clothes and Fashion 1: Stylish, Sassy and Sustainable	322
Clothes and Fashion 2: When Buying New Clothes is a Good Idea	336
The Law 1: Get Up Again, Over and Over	347
Energy 7: The Environmentally Friendly Hearth	361
Energy 8: If I Were a Rich Man: Heat Pumps & All That Jazz	364
Nature 1: Love, Peace and a Nightingale	370
Postscript: And Where From Here?	394
Appendix 1: Further Reading	397
Appendix 2: Further Cooking: Fifteen Favourite Plant-Based Cookbooks for Flavoursome Inspiration	400
Appendix 3: Ten Environmental Charities	404
Appendix 4: Twelve Instagram Accounts of People or Organizations Mentioned in this Book	407
Appendix 5: Ten More Environmental Instagram Accounts to Follow (+ Mine)	408
Acknowledgements	409
Endnotes	413

Introduction

What I'm doing is this. I'm looking for every single way that I can help the planet. How we live and work, travel, shop, eat, drink, dress, vote, play, volunteer, bank – everything.

We all want to do what we can to make a difference. But it's a bit overwhelming – and we're busy. We either have too many school or university exams, too many bills or way too many humans in our lives. We don't have many hours in the day spare and few of us are getting enough sleep. So, as with my previous books, I've done a little research on your behalf. All you have to do is read and use any of the ideas that you like.

And life is short. So joy is vital. There is a lot of overwhelm around this subject which doesn't help. Where do you start? The good news is that there is much that we can do. So much. Let's enjoy changing the way we live and make choices that enhance life for ourselves and for others. Anything less is a disservice to our one brief life and to our beautiful planet.

I've written this in a way that will suit those of us who read but now spend more time on social media. It has short bits and long bits. It will suit you if you have small children and so have long since given up hope of being able to sit down for more than ten minutes. Or if you're intention-rich but time-poor. It has stories, reflections, ideas and, most importantly, every single solution I could find. This is the single-minded focus: solutions. I hope it will be funny in parts and that, for those of us determined to do what we can to love our planet better, it will be useful. It will be good company if you just want to read two pages at the beginning or end of the day. It's not a linear journey because everything connects to everything. Instead, the sections build together, like a jigsaw, to make a picture and a plan.

Please read with a pen or a pencil. Be sure that your writing implement is made of metal or wood and not of plastic. No one needs a home with 32 plastic biros. Those days are gone. You'll see. It will become a habit. Anyway, with your beloved non-plastic writing implement, please feel free to underline. To cross things out. To find me wrong. To find something useful and draw a star in the margin. You even have my permission to turn over the corners of pages. Please steal my ideas. Make them better.

So that's the 'joyful' bit. As to being an environmentalist – who or what is one of those? Wangari Maathai? Jane Goodall? David Attenborough? Greta Thunberg? The dictionary defines an environmentalist as 'a person who seeks to improve the quality of the natural environment and to protect it from harmful human activity.' If you're holding this book you probably recognized yourself in the title. That's you. That's me. That's all the sane ones who live on this earth.

And why do we need to practise without preaching? Surely I'm preaching here, you ask? Ha – well spotted. Yes, I am, but there is a difference between solicited advice and unsolicited advice. If we preach to others by giving advice that people haven't asked for or telling others what they should and shouldn't be doing, we are likely to lose friends and alienate people. Most importantly, because humans are often perverse, (lots of overgrown teenagers around), this is unlikely to create what we need – which is major change in the way that we think and live. What we can do is put our own house fully in order – our own life – and this book has some ideas as to how you can do that.

Don't be discouraged by people saying your actions are just a drop in the ocean. What's an ocean except millions upon millions of drops? And if you've ever seen a stalactite or a stalagmite, you'll know small drops can produce spectacular results.

You are an environmentalist. One who wants to be part of the solution. We are many. We are millions.

PLASTIC 1
Dangerous Woman in Whole Foods Without Fork

Cycling through Battersea today I decide to treat myself to a fresh salad for lunch in Whole Foods. Sounds absurdly middle class, doesn't it? But it's honestly the only place around here where you can get a salad with more than lettuce and tomatoes.

I weave past the well-dressed women in search of 'just ripe' avocados and through to the café. The options all look healthy and nourishing and I ask for three lush tasty heaps of gorgeousness on my plate. Then I stare wearily at the orange juice in plastic bottles.

'Is there any way that I could get a juice in a normal mug?'

'No, madam, we only sell them in these bottles now.'

'But you have mugs right there and juice right there?' I point at the close proximity of the two items I need but it's impossible for them to put juice in a mug, apparently.

'Sorry, madam. Bottles only.'

'But I'd use that bottle for 2 minutes, then it will take 400 years to biodegrade.'

'Sorry, madam.'

Sigh. OK – no juice for me.

I point to my salad. 'Can I have it served on a real plate as I'm eating in, not taking away?' I ask.

'Yes.'

'And can I have a metal fork please?'

He is just ringing up the perfectly cooked meal. I'm really very hungry – I ran around the park at 8am and it's now 2pm.

'I don't need a serviette, thank you.'

I'm not in the habit of smearing food over my face when I eat so why do I need a paper serviette? Why are we given paper serviettes wherever we go? No one uses them at home. A second later they are thrown away.

'And can I have a metal fork please? You always, very kindly, give me a metal fork.'

'We don't have any metal forks. Customers have to use the forks over there.'

'Yes, I know that you don't usually give out metal forks but I believe that you have a few and I'd be very grateful if you could loan me one please. I don't eat with single-use plastic forks. I think they're just in that drawer?' I point at the drawer where I'm pretty damn sure they have forks. 'I don't want to eat this food with plastic cutlery that I'll use for three minutes and then throw away. Haven't you heard of the campaign to refuse single-use plastic? Am I in an alternative universe?'

'They're biodegradable.'

I stare at the 'biodegradable' plastic fork I'm being offered. It's made of a material which, admittedly, if put into an industrial furnace, would break down. But of course none of this cutlery will end up in an industrial furnace. These would never be recycled in any way at all. All their cutlery is being scooped up by their staff and put into the normal bin. It used to go to China but as China has refused to take any more of Europe's plastic pollution, heaven knows exactly where this fork will end up.

The food looks delicious. The other hungry people behind me are obviously wondering what's wrong with me.

'But you've always been happy to give me a metal fork before …' I plead. 'What's the problem?'

'We don't have any metal forks,' says the person serving, who evidently doesn't care about my seeming madness and is failing to understand the point I'm trying to make.

'Can I see the manager?' I ask, with a mixture of pathetic whinging and hungry desperation.

'Please take a seat and I'll call him.'

I pay for my food and sit down staring at it sulkily.

The manager arrives. 'Can I help you, madam?'

'Yes.' I say in a less-than-happy tone. I do realize I'm now sounding like an extreme radical with too much time on my hands – but how else do we get huge stores to change their ways other than by pressure from customers? I take a deep breath … 'I've been one of your loyal customers for about ten years. I'm interested in protecting the planet and so I don't take your "biodegradable" forks because I can tell you that if I put one of these in my compost bin it will come out ten years from now and look exactly like this. You have to understand that "biodegradable" is relative and this is about as non-biodegradable as biodegradable gets.'

He is politely silent. It's possible I'm not being clear. He glances at the offending forks. 'How can I help you?'

'Well – other than changing your policy – I just want a metal fork. You've always been able to give me a metal fork before but today your staff member was unable to find one despite the fact that he didn't look.'

'Of course, madam – I'll get you a fork.'

'Thank you.'

I feel victorious although I'm sorry that I'd had to make a fuss. Off he goes. The hot parts of my meal get a little colder. People around me glance in my direction while eating their lunches with their perfectly functioning non-biodegradable cutlery. I read posters on the wall about what an environmentally conscious company they are.

About ten minutes later (perhaps he hoped I would crack and eat the food with my fingers) he comes back.

'I'm sorry – we don't have any metal forks.'

What planet is this?

'You make and serve food here and you don't have any forks? Not one?'

'It would appear not'

'You had forks last time I was here.'

'It seems they have been removed.'

'What? Why?'

'Health and safety.'

He really says this. I'm not making any of this up.

'Safety? Nothing to do with the fact that you don't have to spend zero point one seconds washing up forks then?'

'No, madam. It's a health and safety measure.'

Strangely, a range of sharp metal knives are clearly visible – presumably the plastic isn't good for cutting up meat. So, they give out metal knives. Knives. But not forks.

'Health and safety? You mean in case I attack someone with a FORK?' For some reason (maybe it was the hunger – maybe it was the sight of the other customers throwing away the plastic bottles, plastic trays, plastic lined cups, all of it) – I actually lose my temper with the poor man.

'CAN YOU TELL ME HOW MANY TERRORIST ATTACKS HAVE OCCURRED IN WHOLE FOODS WHERE PEOPLE HAVE BEEN ATTACKED BY CUSTOMERS WITH FORKS? I mean – do you actually believe any of this stuff you are saying to me? "Health and safety"? Do you realize we have a planet to save? Do you imagine that someone else is going to do it? What hope is there if I can't even eat lunch with a metal fork? Do I look like a terrorist to you?'

He does look pretty terrified actually.

'I'd like to return the food. This is the most ridiculous thing I've heard! "Health and safety"? Why are the knives still here? If I were going to attack someone, I'd choose a knife! Wouldn't you?'

I hear laughing behind me and look around. It's Jeremy Clarkson – the man who famously got into lots of trouble for losing his temper is very much enjoying seeing me lose mine. At least, I'm pretty sure it's him.

'Dangerous Woman in Whole Foods with Fork!' he says. Momentarily distracted from my rage, I just have time to consider whether to risk pointing out the obvious and inform him – as if he may not know – 'You're Jeremy Clarkson.' Or risk upsetting some other man who surely must be informed on a daily basis that he's a doppelgänger. The manager is speaking to me. I stop caring about Clarkson/doppelgänger man.

'Can we offer you something else? You could eat the wraps with your hands?'

I'm beyond compromise at this point.

'No, I don't want something else,' I rant. 'I'd like a full refund.' Now I feel sorry for the rejected food which they will no doubt throw away. I hate waste and this makes me even more angry. 'And I won't be eating in here again.' We can imagine how sorry they will be about that. 'And I'd like to say I'm really disappointed with Whole Foods – I mean this used to be a store that genuinely cared about the planet more than some others. You are claiming to support organic farmers – look at the posters on your walls. But when it comes to a genuine forward-looking environmental policy – you obviously don't have one.'

'The rules come from head office in the USA,' the manager ventures.

'Ah, yes – the "we can't do anything" line. We have no influence. Let's just wait for someone else to save the oceans shall we? Does Whole Foods have an environmental policy or not?'

He points half-heartedly at the wooden drink stirrers.

Clarkson-man goes on chuckling in the background. I glower at him. Then go back to behaving appallingly with the poor manager …

'Thank you. Very much. For the money back. You won't be seeing me again. You have one less customer. And I'll be sure to tell everyone I know.' I try to sound powerful and influential. I don't think he's anticipating the international Whole Foods chain closing.

As I leave I go on ranting at the woman at the till. 'Did you see that? They wouldn't even give me a fork so I could eat my flipping lunch! Poor planet. It's ridiculous.'

'But if you want to save the planet there is a simple solution,' she says.

'There is?' I ask, visualizing oceans full of floating plastic debris. 'Please – tell me!'

'You have to bring your own metal fork.'

Duh – of course.

PLASTIC 2
Ten New Plastic Products Sold to Us on Social Media This Year

Oil companies have spent $180bn in the last seven years finding new products that we need to buy.[1] Here are ten of the items they have created and promoted, all of which have arrived in my feed this year on social media in an attempt to get me to hit the 'buy' button.

1. A plastic pram for pushing a watermelon – includes chiller to keep your melon cool. £130
2. A plastic contraption for hard boiling eggs without the shell. You take the eggs out of their shells, put them into an egg-shaped container which you then place in boiling water. £14
3. A seat with an attached step that fits over your toilet so that your toddler can climb up the one step to sit on the loo safely. Comes in pastel pink or pastel blue so your child knows what sex he or she is. £26.58

TEN NEW PLASTIC PRODUCTS SOLD TO US

4. A doormat made of plastic which is apparently superior to a normal doormat. (However, unlike traditional doormats, will never biodegrade.) £90.72
5. A plastic Dust Daddy attachment for a vacuum cleaner with strands to suck up 'delicate objects' – like leaves and dust on books (as if a normal attachment wouldn't do the job). £10.99
6. A plastic egg to hold your soap powder in your washing machine. Marketed as an 'eco egg' – although quite what is 'eco' about an unnecessary plastic egg isn't clear. £9.82. Buy a different one for your dryer too. £9.99
7. A 'mascara shield': a piece of plastic shaped to go over your eye so that you don't get mascara on your eyelid when you are applying it. £6.89
8. An eyeglass cleaner on a plastic stick (cleans both sides of your glasses at once) to save you using a cloth. £8.34
9. A 'toothbrush of the future' that is made entirely of thermoplastic. Says it's recyclable – it certainly isn't biodegradable. $12 in the US.
10. An intimate shaping tool – a piece of plastic that you put between your legs to shave your pubic hair into a heart shape, a downward arrow or an exclamation mark. £14.95

EARTH 1
Plant Your Trees

Relax, put your feet up, this is a longer bit ...

I've come to Inverness to volunteer with a charity called Trees for Life. They are recreating the ancient forests of Scotland – the Caledonian Forests – and their work is a joyful miracle.

The forests of Scotland have a problem. Well, they have several problems – namely a large number of Scots, Americans and rather wealthy, privileged English. (I dare say that even the Welsh and the Irish are not entirely blameless.) The Scots, first of all, are very proud of their deer. In fact, if you put 'Scotland' into Ecosia images, it's highly likely that on page 1 you'll find an image of a stag in a desolate landscape. The beautiful stag standing on a barren, treeless hillside is the classic postcard image of some people's idea of what Scotland looks like. But why is the land so bare? Why is it like a desert? Because deer eat. Everything. As do their friends, the woolly ruminants from Mesopotamia, otherwise known as sheep. The writer George Monbiot describes sheep in that way, explaining that, while they are not even indigenous to the UK, they have done more damage to our countryside than all the building work that has ever taken place. The deer and the sheep, you see, eat baby trees and almost everything else so nothing can grow. It's a disaster.

We don't think of sheep or deer as enemies because if we are British we are used to idyllic pictures of stags, deer and fawns and herds of shaggy sheep dotted across lush green fields like harmless fluffy clouds in a blue sky. But they are not so harmless. When you have ruminants on the land, quite understandably they want to eat. Of course, there is nothing wrong with deer and sheep; it's the excessively high numbers of them that cause the problem. And why are there excessively high numbers? Pesky humans and their desire for money.

Many people, for reasons that are totally incomprehensible to me, seem to find their life is enhanced by causing pain to innocent creatures. It costs about £700 to shoot a stag. More if your ego requires that you take the antlers home with you to put on your wall. (Seriously, people want to do this.) For lots of men this is their idea of a good day out (to be fair it's almost exclusively men), and so if you're a Scottish landowner, the more oversized stags you have, the richer you will become. The result is massive overbreeding. There are way too many deer in Scotland, which leads to barren hillsides and no trees – resulting in the breakdown in the quality of the earth, flooding, no birds, no other wildlife (except grouse, but that's another story – they are also bred for profit). Ultimately, ruining nature as we should all be able to enjoy her. Her – yes, let's call nature 'her'. Maybe that's the influence of the French language on me – la nature – or maybe it's something a little more primal. But let's not go there.

As for the woolly ruminants ... farmers are not allowed to keep land that is 'doing nothing'. If they have a piece of land and put sheep on it, they can receive subsidies because it's called 'farming'.[2] Another rule that I'm totally unable to understand is that you are not allowed more than an absolute minimum of trees on the land. Because if there are trees on the land then it's not farmed. Something like that.

Basically, what I'm saying is that landowners can make money if they have sheep or deer on their land. And because deer and sheep want to eat, the trees can't have any babies that will grow even into teenage trees. It's quite sad. Any time a tiny sapling peeks its head up through the earth, something comes along and eats it. This has led to what is known as 'geriatric forests': forests in which all the trees are fully grown, old or dying or dead. No saplings growing beside them.

But – your narrator announces proudly – the genius beings at Trees for Life have come up with a very simple solution: fences. If you take an area around a forest and put up a fence to keep out our four-legged friends, in no time at all the seeds from nearby trees land on the ground and some of them take root. This is obviously annoying for the deer who have to stare at yummy saplings from the wrong side of what has to be quite a high fence. But it's very good for the saplings. And the wild flowers. And the ground-cover plants. And the scrub which protects the young trees. And the birds. And the insect life. And the quality of the earth.

So that's one thing they do – they put up fences and allow the forests to recover. But they also do something still more radical. They befriend Scottish landowners. They are nice to them. And the landowners have to admit that what they are doing looks like a good idea and so (biting their knuckles and imagining three less stags) the landowners allow Trees for Life to fence off pieces of land in the areas they own and plant trees on them. Brilliant, no?

Also, and perhaps more exciting still, a very wealthy American person whose name I don't know but who didn't spend his or her spare time shooting animals, gave this small Scottish charity a substantial amount of money to buy their own land. This was a dream come true for them as it's a very large piece of land (basically an entire hillside) and they can plant exactly the trees that would be growing there were it not for man's incessant desire for profit.

That's why I'm here. I'm planting trees. It's knackering.

The first evening, we are out in the woods at night. Our torches are still in our backpacks as we can see by the light of the moon.

Our pack leader crushes something between her hands and holds it under my inexperienced nose.

'What does this smell remind you of?'

'Cheap bathroom products?'

She looks mildly despairing.

'And what do they put into bathroom products?'

'Lots of chemicals? Artificial reproductions of smells?'

'I suppose you're right, but – think now. What do they put into disinfectant? What are the smells they recreate based on? You know – the real smells? From nature?'

I sniffed her cupped hand again.

'It's vaguely familiar but it doesn't smell like Dettol.'

'It's pine! Look! Pine needles.'

Ah, yes, pine. Cheap bathroom products. To me, a city dweller, it's all back to front. I'm not sure I've ever smelled pine needles like this direct from a tree. (Christmas trees smell different.) And, I would guess, I'm not the only one who would associate the smell of pine with bathrooms instead of trees.

*

We get up at 7am, discuss who is using the shared showers first, grab breakfast and fill our non-plastic sandwich boxes with a vegetarian lunch. All the food here is vegetarian. This makes me very happy. There are only two meat eaters among the volunteers, which makes a refreshing change. Vegetarians have been viewed as the eccentric odd ones out for so many years that it's nice to be in the majority. We fill flasks and put on our layers of clothing which have to be suitable for walking up what feels like an almost vertical Scottish hillside, planting trees in full sunshine or spending parts of the day in the rain. The gaiters are my favourite clothing item. I've never owned gaiters before this week. They are worn over the waterproof trousers and walking boots to keep out the deer ticks.

The ten of us pile into a van and are driven half way to the planting site, as far as the road goes. Then out we scramble, shovels and planting bags in our hands, and off up the hill we go. The youngest of us volunteers is 19 and male – he goes up the hill at the speed of a young mountain goat. Fortunately for me there are also some older volunteers. Pensioners, thank goodness.

Being an author mainly involves sitting on your bum where, on a good day, some kind soul will bring you tea. Running around the park occasionally must make me vaguely healthy. Because I can do this and have never smoked I consider myself reasonably fit. Apparently not. Maybe it's the shock of the fresh air, but I'm fantastically tired from the first walk up the hill which they claim takes 30 minutes and I would swear takes an hour. Boggy, it also is. Very boggy.

We reach a high fence and go through a gate. Good luck, I think, as we squelch our way over land that belongs to

Trees for Life, to any deer that tries to get over or under that. Everywhere are little mounds of earth that have been turned over by some kind of machine (heaven knows how it got up here – maybe they flew it in) and on each mound are newly planted saplings. It's strangely, mystically beautiful – like a promise that a future will come, after all, for our precious planet.

We arrive at the area that we are going to be planting and hundreds of bags of tiny trees are sitting there waiting for us. There are saplings of downy birch, rowan, willow, Scots pine. I can also see holly – at least there's one leaf that we can all recognize. We're given a large yellow planting bag with as many batches of 15 downy birch trees as we feel like carrying.

Downy birch (which, to my untrained eye looks the same as silver birch) is a 'pioneer species', which means that it can grow on poor-quality ground and, by growing there, will improve the soil, making it easier for the seeds of other trees to take root successfully. The trees we are planting today will take 25 years to reach seed-bearing age, but no one here is in a hurry.

'Who wants mycorrhizal fungi?' asks one of our group leaders. We all do. We're going to give a spoonful of mixture, combined with a little charcoal, to each tree to help it grow.

'Select your spot, push your planting spade in like this, make a hole that's the same size as the root, put some of your mixture in the hole, push the tree down firmly so that if a baby deer gets in they can't pull the tree up, and that's it.'

'Is that all the attention the tree gets for the rest of its life?' someone asks.

'Yes it is. If you'd like to you can dedicate the tree to someone.'

'How would we do that?'

'Just think of the person as you are planting the tree. Simple as that. But not for all the trees or it will slow you down.'

'OK.'

'And you can give the tree a kiss if you like. This charity was born at Findhorn and they always say "plant with love". Come back here when you need more trees.'

And that's us. We're set for a few days on the hillside.

If you want to get fit and feel good, this is the trip you need. Exhausting? Yes. But transformative. It's like bringing life back to a desert. It's exactly that. And they don't even need to remind you to plant with love because there is something about these young trees that would draw tenderness from the hardest of hearts.

This hillside is barren. There is no shelter, no protection from the biting wind or the driving rain. This is Scotland and we're in the Highlands. The saplings look so fragile. They are fragile. They have each been grown lovingly from seed and when we push them into the cold ground they are between 6 and 9 inches high. We give them one meal in their planting hole, one kiss and then they are on their own. But they have two friends each. We're planting them in groups of three: two downy birch and one rowan per overturned area. Trees, as we now know, have ways of communicating with each other and supporting each other under the ground. Seriously, they do. If nutrients are scarce, they share them; if danger arrives, they warn each other. Just don't ask me how – but they do. So, the trees have two friends each. As they grow they will protect each other.

We dig another hole, add mixture, push in a tree, fill up any air gaps with earth and repeat. Four hours later we stop for hot coffee and tomato sandwiches. Then we plant from 2pm till 5pm. This is what I call a 'Zen job' – it requires lots of action and very little thought, other than the occasional, 'Would this tree be better higher up the mound or down a bit? or Is that pushed in firmly enough?' Apart from having to watch where you put every footstep, there is no room to think. No thoughts of past broken hearts or future anxieties. Each tree is far more important. And so, with each of us fully noticing the miracle, with no breath to talk but with our eyes and our hearts open – we plant another tree and another until, with ten of us working for a day, another thousand trees are planted and we are walking carefully back down the hillside, as happy and satisfied as a bunch of scraggly volunteers can be.

Is it possible to fall in love with trees?

The evenings are cosy here. Computers and screens are banned in the communal living spaces so, radically, we talk to each other and play games. There is an open fire. There is homemade vegetarian food which we take turns to cook. There is a possible flirtation between two of the younger volunteers, one older friend grieves his late wife and another volunteer has a wife at home nearing the end of her first pregnancy. She has made him promise to finish the week – we're excited for him. The youngest of us is 19 and the oldest 72. We are gay and straight and no one cares. We talk about trees – passionately. I have never been with a group of people before who discuss the qualities of aspen or can list different species of willow or know the Latin names of plants.

'Learning the Latin names was part of my training as a focalizer for these groups.'

'A focalizer? What a wonderful word. It has "folk" in it.'

'It just means someone who hold the focus for the group.'

Our focalizer smiles as I express my admiration of her knowledge of *Salix aurita* – eared willow – which we'd been planting that day.

I have never listened to people before that care so much about trees and the future of our planet. It's so inspiring and humbling at the same time. I sit down to write a list of the trees that I think I can identify by sight. Can I think of ten?

Here are the trees that I think I can identify by sight at the beginning of this project:

Oak
Horse chestnut
Sweet chestnut
Silver birch
Japanese cherry (in spring)
Copper beech
Holly
Weeping willow
Sycamore

And I recognize a Lombardy poplar as there were some behind the house I used to live in before some moron cut them down. I remember, I cried.

But I couldn't tell a beech from an elm or a rowan from an ash, and even though I've included the sycamore there is a good chance I may confuse one with a maple because I know that the leaves are a similar shape. The maple leaf is the one you'd recognize from the Canadian flag.

And then the ability to give names to a small group of trees hardly gives me any knowledge of them. Labelling items in nature brings us no closer to them, does it?

PLANT YOUR TREES

What else do I know? The seed of the oak is an acorn, the seed of the horse chestnut we affectionately call a conker and the seed of the sweet chestnut we can take home and eat. I know that a yew tree is often found in a churchyard and that it can live for a thousand years. But if I did see a yew tree, I wouldn't be sure that it was one or just another very old-looking tree. I don't know the difference between pines and spruces. I'm never sure whether a London plane tree is or isn't a London plane tree, and although I have an idea which houseplants are best for air quality in the home, I have no idea which trees are best for us or for our gardens.

Here, people have conversations about the fact that *Juniperus communis* is intolerant of shade. This delights me utterly. Because most of us can't identify even the trees in our own parks. We don't know which trees are good for the bees. We plant for appearance rather than because a tree has berries that will attract birds. It's sad how little we know. We don't question why a forest, when we walk through it, has only geriatric trees with no young growth. Or at least I never have. Have you?

How many trees can you name? Perhaps you'd like to play with me? Put this book aside, grab some paper and write down the names of all the trees that you could identify along with everything you know about them. I hope your list is longer than mine.

But I have to go to bed now as we are getting up very early and those hills we are walking up are steep.

And as I walk up the hill I'm thinking about my foot slipping on a rock. I'm in agony and we are miles from base and no one can carry me back to the van on a stretcher because it's too steep and I'm crying and shouting 'fuck' because it hurts so much and they have to call the local rescue helicopter to get me off the mountain. And I ruin the day for everyone and it's very expensive and I'm in too much pain to enjoy the helicopter ride and worst of all I can't plant any more trees …

No, that hasn't happened – thank God – but I wanted to share with you what I'm thinking about as I walk up the hill. The scene that is playing out in my mind, in detail, as I walk very, very carefully on the rocky, boggy ground. I have even thought of the colour of the jackets of the paramedics. For some reason they are grey in my imagination. The stretcher is grey too. The team is mixed, two men and one woman. They all have very soothing Scottish accents but, all the same, I don't want to meet them – this professional team that my unconscious mind is summoning. Why do our minds do this? The reality is that the sky is a translucent blue today. My boots are sturdy, my gaiters keep out ticks and the sun is shining.

'Walk relaxed,' says one of the team leaders. I appreciate the mental reframe.

'Thank you.' I breathe better. I was doing 'walking carefully to avoid injury' – so putting all my focus on possible injury.

'Better to walk relaxed,' she says. 'Trust your step.'

Like all good group leaders, they lead us up the hill at the speed of the slowest walkers. I wonder how many days of this it would take before I became accustomed to this much pure oxygen and started to get fit. Do you ever imagine what your body would feel like if you were super fit? If we all lived lives where we did enough outdoor exercise every day in fresh, unpolluted air? And not even exercise, just daily contact with wild nature? What would happen to our physical and mental health?

I feel so happy standing on this hill. I guess anyone who has just walked up a hill for an hour and reached the stopping point is happy. But not happy in a way that I would if I was just here to take a photo and walk down again. I'll be happy here all day. In our own small way we are restoring the great Caledonian Forest.

This morning, birch, rowan and willow are distributed. Our work is set for the day and our minds can tune into a zone of beautiful simplicity again. Select a good spot for a sapling, dig a hole, add the charcoal and mycorrhizal fungi, unwrap a sapling from the pack of 15 saplings, kiss the sapling with all your heart, put it in the hole, fill the edges with the earth you've just taken out so there are no air pockets, press down firmly, say 'good luck' and step backwards carefully to avoid any need for that paramedic helicopter.

The rowan, or mountain ash, is incredibly pretty – the leaves are just turning yellow, golden orange and red as if the tree is having its own little autumn. I've done just a little bit of research about the trees I'm planting and so I know that in about 15 years the rowan will be covered in bright-red berries that will feed the birds every year. Each tree, if it takes well and doesn't get eaten by a baby deer sneaking under the fence, can live for up to 200 years.

The lack of cover is a problem working all day on a hillside. If you want to pee, your options are either to miss your chance to plant more saplings by going for a very long walk, or throw dignity to the wind and pee where everyone can see you. The second problem with taking a pee is the ticks. You really don't want to squat down in the grass and have a tick attach itself to places that you don't want to be showing to a fellow volunteer later. So, I sit on a rock reasonably high up the hill and see a man glance up the hill and then look down again hastily. Not caring too much about moments like this is one of the requirements of the job. I drink a little coffee poured into the cup from my Thermos flask. Back to planting.

If I step back and my foot sinks into a bog then it's time to plant willow. The willow likes the boggy bits. It's not weeping willow, which before this week was the only one of the more than a hundred species of willow that I could have identified. This has a darker green and rounder leaf, and it likes wet places. So, when it's too boggy to plant downy birch or rowan, we plant eared willow. Not too wet – if the hole you dig fills with water, it's too wet. Not too dry – if a mound on top of a wet area has dried and it's not soggy, it's too dry. It surprised me that we didn't have anything to water our trees immediately after planting. In gardens I've known I've never planted anything without watering it in. Where there was a little puddle close to the planting, I scooped water and watered the trees in. But we were not instructed to do this. The clods of earth the trees are in are moist, and this is the Scottish Highlands after all. I guess that it will rain soon enough.

This land is unpretentious – it doesn't mess about. It has an honesty about it. You can see why the Highlanders have a reputation for being gruff and honest. There can be no deception living on land like this. Quite simply, you can see everything. Dundreggan where I'm standing is in Glen Moriston. Trees for Life bought the 10,000-acre area of land in 2008 and they have been planting trees and encouraging native regeneration since then. They are aiming, according to their website, to: 'Develop Dundreggan into one of Scotland's finest native woodlands and create a link between Glen Moriston and Glen Affric.' Which will be a haven for wildlife and protected, they have good reason to hope, for generations to come.

The bare hills with the occasional crow passing over and deer in the distance has a certain beauty ... but if I close my eyes in the sunshine I can imagine trees here. And with the trees the shrubs; the flowers; the butterflies; the many species of insects feeding from the wood; the enriched bird life feeding from the insects, maybe, as they hope; wild boar; beavers; badgers; pine martins; lynx and even wolves roaming freely on the hillside. The deer, the sheep, the rabbits, hare and grouse would still be here, of course, but there would be less of them.

And people. There would be people because when there is a rich natural environment of this kind there is always eco-tourism, as people are keen to get away from our polluted cities and experience nature as it can be. There would be ramblers and, in my vision at least, more people shooting with cameras and less people shooting with guns. There would be no one taking home antlers to mount above their fireplaces, but there would be people keen to learn which herbs and plants could be freely foraged and taken home for a salad.

Instead of children in primary schools having to learn about how our forests are being destroyed, they could learn about how they were being destroyed until we began protecting them again.

I'm disturbed from my reverie. One of my fellow volunteers – Ian, from Edinburgh, where he usually works in an office – is waving and pointing frantically. I glance up to where he's signalling and there is a huge bird. Of course, I have no idea what kind of bird it is. I'm from London. But it looks graceful and like a bird of prey of some kind. I watch it for three or four minutes until it swoops out of sight.

Ian treads his way up the hill carefully to get more downy birch.

'Did you see that?'

'Yes. What was it?'

'You don't know what that bird is?'

'I have no idea.'

'I'll give you a clue. It's over two metres across.'

'No – really, I've no idea.'

'It's big and rare. Even in these parts. Come on now!'

'I've no idea. Some type of bird of prey?'

'It's an eagle, woman from London. It's a golden eagle.'

The only place I had ever seen an eagle before was in captivity at the International Centre for Birds of Prey in Gloucestershire. But here – already, as a sure sign of hope – is a golden eagle in the wild.

In the middle of the week we had a day off to work in the nurseries. The tree nurseries grow over 60,000 native trees every year. This blows my city-dwelling mind a little bit. We have a celebration in London if we plant one tree. And in some cities, the local councils have become famous for the ability to cut down trees (Sheffield, I'm thinking of you. Plymouth, are you reading this?) – usually at dawn before the tree-loving local residents have time to form a human ring around the trees and beg and cry for mercy. But let's not think about such insanity.

Here there are sane humans. They collect the seeds themselves. They send out volunteers to collect seeds from local trees. This is so that the trees are of the right genetic stock to survive and thrive in local conditions and have the right immunity to disease. Just as humans that live on high mountains (in Tibet or South America, for example) have adapted to breathe at altitudes that would be very difficult for the rest of us, and developed an immunity from the local bugs, so trees build up immunity to tree diseases in certain areas of the country.

Today we are learning about aspen. It's so beautiful. Just put 'aspen images' into Ecosia and have a look at them. Oh, Ecosia – I may need to explain. Here is a very easy thing that you can do today to love the planet a bit more. Change your search engine to Ecosia. It works exactly the same as Google and other search engines – except they plant trees. And it's really easy to change as it only takes two or three clicks. For every 45 searches you do, and you'll be surprised how quickly they add up, they plant a tree. They have planted over 39 million trees and if you go to their info page you can have the satisfaction of seeing the figure going up right before your eyes. It makes you feel good every time you do an internet search. So that's your new search engine sorted, I hope.

But back to aspen. Aspen is very beautiful and has an unusual reproductive system. It doesn't flower and it doesn't have seeds. Also, you can't grow them from stem cuttings as you can with, say, elm or birch. Aspen spreads by sending up new stock from its roots. So, in early spring, volunteers bring in root cuttings and look after them in their polytunnels. Each root cutting has to be planted individually, so it's expensive and labour-intensive, but Trees for Life have been working with aspen since 1991 and can now produce up to 3,000 young aspen a year.

There is something so full of hope and joy about polytunnels full of baby trees. There is beauty all around.

Following our little tour of the aspen project we are set to work gathering rowan seeds. This is a job that can be done sitting down, as a previous batch of volunteers had been out collecting the berries. Our job is to crush the berries in buckets of water to create a mulch which could then be sieved and the seeds collected and planted later. Imagine crushing a bucket of grapes except instead of wanting the wine we want the grape pips. We are given gloves and we sit and squash each berry.

'You need your bucket to resemble a bucket of vomit,' says our leader.

It's cheerful work.

'Do you want an explanation of an acorn?' a man asks. A volunteer that I hadn't met.

'I do.' I want to learn everything I can.

'In a nutshell, it's an oak tree.'

I look confused. 'Is that a joke or a miracle?'

He thought about it. 'Now that you ask, I'm not sure myself.' An oak tree, inside an acorn, inside a nutshell. I mean how does that happen?

'Is it true that oaks support more wildlife than any other trees?'

'That's what they say.'

'Then why aren't we planting any this week? I'd love to plant an oak tree.'

'Some will be planted in the area where you're planting this week but it's a bit high for them up there. They prefer a lower position. If you can find a sheltered place, it's OK.'

'There aren't many sheltered positions on the bare hillside.'

'Exactly, so we don't put them there. If they manage to establish themselves up there it's great but they won't grow to the huge height that they can if we plant them lower down.'

'There is something so magical about an oak tree – I'd almost say mystical.'

'It's considered a holy tree,' one of the women says. 'Oak, ash and thorn are supposed to be the holy trees.'

A local pensioner who has joined us for the day pipes up with his rich Scottish accent. 'All trees are holy in my book.'

'But acorns are said to have many magical qualities,' replies our woman mystic. 'You're supposed to be able to attract a member of the opposite sex if you wear them. And they're symbols of fertility and creativity too. I guess it's because so great a tree comes from so small an acorn. There is even the belief that if you catch a falling oak leaf you'll be free of illness for all the cold winter months.'

'I'm not so sure about that,' says the pensioner. 'But the acorns make good food for deer. And for wild boar, once we're allowed to have them.'

I love listening to this conversation. I have to stop squashing rowan berries to take notes so that I can remember all this later.

'Don't underestimate the magical powers of the oak,' she says. 'They say that if you listen to the leaves blowing in the wind in full summer, you can tell the future.'

'And even if you couldn't, you'd feel better listening to the sound of the leaves.'

'That you would. Everyone loves an oak,' says the Scottish pensioner. 'I've asked them to plant one over wherever they lay me.'

We are slightly startled by this.

'Do you find that surprising? That I would have asked for that?' he asks.

I suppose not, I think, not sure whether this is sad or not. I've only known the old gentleman for an hour but I'm already sad at the thought of him lying dead in the ground. On the other hand, I guess him turning into an oak tree is better than being cremated. Then he can whisper people's futures to them in the wind.

'Pass the berry buckets around now,' says our leader, shocking me back into the present moment.

I had a new bucket. Full of bright-red rowan berries to squash.

'How did you come to be here?' I asked my new neighbour. 'Are you staff?'

'I am,' said the young man. 'I'm a trainee, so sort of staff. Nick, pleased to meet you.'

'How did you end up here?'

'It's a long story.'

'We have a lot of berries to crush and I like stories.'

'OK. I did a degree at Bath Spa University in History and Cultural Studies.'

'Useful for berry crushing.'

'It took a long time to get from there to here.'

'So you didn't come here straight from uni then?'

'No, I got a job working at the Roman Baths.'

'In Bath?'

'Yes. I worked there for a while. Then I worked at the National Portrait Gallery in London. And I worked in telephone banking for Lloyds, in a letting agency, at NFU Mutual as a mutual insurance claims handler, as a gardener for Bristol Council, driving cars for car auctions, and I've done apple picking and juice making in Somerset.'

'A serious commitment to discovering what you want to do.'

'And at the same time I volunteered at a permaculture small holding, an apple orchard, a permaculture allotment scheme, a biodynamic farm and veg box scheme and visited loads of permaculture and biodynamic farms and gardens while taking an introduction to permaculture course. The environment is my real passion.'

'An excellent passion.'

'But of course I needed money, as we all do. So I couldn't afford to give up working and just become a full-time environmentalist, and as I still haven't paid off my student debts from my original degree, I certainly couldn't pay to retrain.'

'I know so many people in this position.'

But then I heard about this Trees for Life traineeship funded by the Heritage Lottery Fund. It allows young local people, career-changers or women commonly underrepresented in male-dominated roles such as deer stalking, to gain training in the skills required for restoring forest cover to Scotland.'

'So you didn't have to be qualified to do this?'

'The reverse. I wouldn't have been accepted if I had qualifications in horticulture or any environmental qualifications.'

'That seems a bit unfair on the people that went to uni for three or four years to study. They wouldn't be able to apply for the place you have.'

'Hopefully they'd be able to get paid placements. This is entry level. They should learn on their courses everything that I'm learning here. As trainees we're not paid to be here but we do receive a small bursary and free accommodation. This money is from the Heritage Lottery Fund.

'I see. So you won this place. That's fabulous. Why do you think they chose you?'

'I was lucky, I guess. But I do have a deep commitment to restoring our landscape. I had been volunteering for about six years. I hope they could see that I'm genuine. I want to do something deeply meaningful and lasting. I told them that co-creating a healthy, happy ecosystem would bring happiness and fulfilment to my life.'

'Co-creating?'

'With nature that means.'

'I see. So does it bring those benefits? Or do you secretly hate being in this remote area of Scotland when it's freezing cold and pouring with rain? Do you long for the city-based pleasures of the National Portrait Gallery, theatres, cinemas and the art galleries?'

'I don't dislike the National Portrait Gallery or any of those things. But this is on a whole other level. I've honestly never been as happy in a job. I love it here. The people are wonderful and it's so beautiful. And I left somewhere very beautiful to be here. Immediately before coming here I was living in south Devon. Those berries are done now. We can go in and have tea.'

We take off our rubber gloves and leave the berry buckets for collection later. As we walk back through the rows of growing trees with the other volunteers, Nick says, 'Honestly, I've enjoyed every minute here. I'm happier than I've ever been in my life. I can recommend restoring healthy ecosystems.'

I love that.

This evening we all have tick hysteria. One person finds a tick on himself. We suspect it to be a deer tick. Ticks should be harmless. They bite you and then fatten themselves up with a little blood. They have been feeding from the blood of other species for 120 million years. The problem is that they can carry Lyme disease (mostly they don't) and the longer they hang around on (or I should really say 'in') your skin, the more likely you are to be infected. So of course once one person finds a tick we all started itching.

The leaders here are super cool about it all, though. They have a special little tool for tick removal and even a microscope so that, once the ticks have been removed from sundry parts of people (mostly arms and legs I'm happy to say) we can have a look at them under the microscope like excited children.

'Have you got one?' we all ask each other. 'I've got two,' some of us reply. It becomes like a competition. We girls all rush off to examine each other's skin. A tick bite is easy to see as there is a circle of red on the skin with a clear black spot in the middle which is the tick. We watch with macabre fascination to see the ticks removed with the special little tool.

By the time we have finished examining every square inch of each other (almost), a total of six ticks have been located on the ten of us.

'We have loads of them here,' one of the permanent volunteers who lives on the site says casually. We even have someone writing a thesis on them. 'If you develop flu-like symptoms a week from now, go to the doctor and they'll give you an antibiotic injection. It's really not a big deal.'

If someone does get Lyme's Disease, though, and it goes undiagnosed and untreated, it can be serious.

But here everyone is so chilled about it. I itch and stare at the centre of red dots on me for black marks. As someone who believes that everything is here for a reason – even ticks – I feel bad killing them. Like everything, they do serve a role in the complex tapestry of life. They are food for guinea fowl and it's even been whispered that perhaps, in some way that we don't understand, they – like Trees for Life – are doing their bit to keep the excess deer population in check. By spreading disease to the deer, they too prevent overgrazing. In 2018 a piece of research found that, in America, there has been an increase in people with allergies to meat as a result of tick bites. Less meat eating is always a good thing. Humans were never intended as their victims but, all the same, I feel bad for those that have to have them removed from their skin and glad that I don't have any myself.

I go back to the bedroom and vacuum it. Thoroughly.

Then, after our veggie dinner, we all sit around the camp fire and drink hot chocolate, toast marshmallows, sing silly songs and stay up too late, seeing as we have to be up at 7am for another full day's planting on the hillside. And we raise a hot chocolate toast to the ticks – and whatever unknown purpose they are serving. If those ancient Indian belief systems that teach the Wheel of Life are right – and we come back as animals and insects higher or lower depending on our actions – you must have done something pretty bad to be born as a tick.

I wake up here far more easily than at home, even though my alarm is set earlier. Maybe it's the fresh air. Maybe it's because I know that today is another day doing something that feels truly worthwhile.

We walk up the hill a little bit more easily today. The body adjusts to exercise. It starts to drizzle. I pull up my hood and think how glad I am that this rain will be soaking into the earth, watering the saplings that we planted only yesterday.

It rains harder. The leaders put up a little tent and we sit in it and drink hot coffee till the shower passes. We watch the streaks across the sky from the grey clouds to the bare hillside.

'Can you imagine this hillside all covered in trees?' someone asks.

Yes, we can. We can imagine it.

The shower passes and we pick up our shovels and start the day's work. After a day of planting, at 5pm we clean the shovels in a stream and walk slowly and carefully back to the van. The oxygen pumps through my veins, nourishing every cell. Every planted tree feels like an offering back to the planet.

The week ends and my wonderful and knowledgeable tree-loving friends and I say goodbye. (The man who had lost his wife had planted a grove for her and was now off to explore Europe, having sold everything he had.) Everyone is going back to whatever they do when not in the Highlands. It has been wonderful. Physically hard, certainly, but so enriching.

I think to myself that even if I were to die next week I would not regret having spent today planting trees. It's not a superficial happiness. I feel like the tree feels – a sense of wellbeing that has roots deep in the earth and hands raised to the light in thanks. It feels – from a woman who has lived most of her life in cities – real. Joyfully real.

PLASTIC 3

The Basics: How to Remove Plastic from Your Home – 36 Steps for Starters

If you're not a starter, you can skip this – or just scan the headings.

Make a game of it: start with the plastic that you only use once ... Begin with the obvious stuff. You know all this already. But apparently some people don't. So just in case ...

1. Plastic pens. No thanks. If you are a stationery lover, this is an instant joy: find an independent stationers that sells real pens. They still exist. Go there. Choose a beautiful pen made of metal. You don't have to spend thousands of pounds buying a top-of-the-range pen. But – if you have a birthday coming up then why not treat yourself to a metal pen that you love? It can certainly be an object of beauty. Mine is metal and fills from an ink pot – not with one of those old levers but with a screw-in section – so I don't even have to buy plastic cartridges. The nib of my beloved fountain pen is

italic so it gives people the impression that I have good handwriting. I don't – but it works miracles compared to plastic biros. You won't lose it because you'll take care of it. Really.

2. Shampoo and conditioner. Commercial shampoos and conditioners ruin your scalp and your hair. They really do. I wouldn't have believed this until I swapped my shampoo in a plastic bottle for a more natural shampoo in a bar form. I didn't expect the shampoo to be very good to be honest with you – but to my amazement it felt mild and lovely on my hair. I found that my hair didn't get greasy and I now only have to wash it about once every ten days. Recently a guest of mine left some commercial shampoo and to avoid the waste I thought I'd use it up. Heaven knows what chemicals it had in it but it sure did strip my hair of all its natural oils. For about a week afterwards my scalp felt dry and uncomfortable. I'm not doing that again. Stick to the bars – and use cold water. The conditioner bars are less satisfactory for me (so far), but there are now more and more alternatives on the market. My present favourite is homemade by a friend. A quick hunt around on Instagram for #shampoobar or #conditionerbar will give you lots of options. You won't be putting chemicals on your head any more either.

3. Razors. Be gone disposable razors forever. Buy yourself one of the beautiful metal ones that lasts a lifetime. They look great and, with their metal blades, they work better too.

4. Soap. In a plastic dispenser? You have to be kidding me. But a bar of soap with your favourite perfume in it will please you every time you use it. Use an old-fashioned

soap dish (to drain the water off the soap) made from china, wood or enamel.

5. Fruit and vegetables. Plastic wrappers around fresh food like bananas, oranges and avocados? Just refuse to buy them. Sometimes you have to be prepared to inconvenience yourself. If you go to a supermarket and the food that you want to eat is wrapped in plastic, take ten minutes out of your day and ask to speak to the manager and explain why you won't be buying the wrapped bananas because you don't eat banana peel. I did this last week and the manager said, 'You are the second person this week who has raised concerns about packaging and whose custom we have lost because of it.' Do they care? Yes, they do. This is capitalism and they care about every purchase. Meanwhile, go on adventures and find all the places locally where you can buy what you need without the plastic. Once you start this it brings the fun back into shopping. You are now an activist. We have a grocer's near where I live where I can buy 90 per cent of my fruit and veg package and waste free. I tell them this is the reason I shop with them and they tell me, 'People keep saying they come here for the waste-free shopping.' You may feel that just your shopping habits won't make a difference. But there are millions of us.

6. Dried products. We have our first waste-free shop locally. You take all your containers with you. They weigh the containers then weigh them again when you've filled them with nuts, lentils, muesli or whatever takes your fancy. To walk to this shop from my house takes 30 minutes but I find it so satisfying and enjoyable to shop there that I'm happy to do the walk. I'm aiming

at zero waste which is an ideal that I don't expect to achieve but it's weirdly satisfying unpacking your weekly shop and having nothing to throw away.

7. Plastic bottles for water. Are you kidding? Tap water is perfectly safe and has to be rigorously tested. Do your own research. Come to your own conclusions, but buy one portable water carrier (not made of plastic) – and use it for the rest of your life. I have a local friend on benefits. She really struggles with money. But she buys two bottles of water in plastic bottles a day. She spends over £820 a year on buying water. I tease her that she must be very rich and point out that I drink tap water every day and don't seem to have died. But there is nothing I can do to persuade her to stop wasting her money and polluting the planet. As you may imagine, I've tried. But please – don't buy water or any carbonated drinks in plastic bottles.

8. Shopping bags. Plastic bags? 'Bags for life'? Insulated extra-thick 'anti-bacterial' plastic bags from supermarkets? Enough already. Don't you agree? If we want to pick up some food from a corner shop and don't have a bag with us, we can't shop. Period. Surely we can learn to carry a thin cotton tote bag that we love? It can't be beyond the wit of humankind to accomplish this. We have a planet to save. And bin liners? Nope. You're not going to have much rubbish by the time we are done anyway.

9. Lip moisturizers. Can now be bought in waste-free shops in cardboard containers. And they are not filled with chemicals either.

10. Toothbrushes. Plastic toothbrushes litter beaches all over the world. Do you really want to buy an item that,

once you've thrown it to that 'away' place, is likely to last hundreds of years after we have ourselves ceased to be? Bamboo is very fast growing and biodegradable. Some bamboo brushes are poor quality and badly made apparently (I've never had one of the bad-quality ones), but the good ones last just as long as the plastic ones. A bamboo toothbrush feels a bit strange at first if you've used a plastic one all your life. It takes about three days until you get used to it.

11. Toothpaste. I just can't buy those plastic tubes any more. You can buy toothpaste in tablet form from Lush in plastic bottles that they take back and reuse (they claim) up to 15 times before they recycle them. Or you can make your own; there are lots of recipes online. Store it in a glass jar. So much more satisfying.

12. Dental floss. If you enter the words 'Dental Lace' into Ecosia you will find dental floss that isn't plastic and even comes in a glass container.

13. Hair brushes, makeup brushes – even toilet brushes. Just don't buy plastic. Alternatives can always be found. But if you also care about animals, beware of anything that says 'pure bristle'. Bristle is most often the hair of a special kind of pig that is factory farmed in Asia. So 'pure bristle' means the same as 'real fur'. Don't buy it and ask the supplier if they know the condition that the pigs are raised in. Sometimes avoiding both animal products and plastic can be a challenge but it's one worth pursuing. Buy wood and enquire carefully as to what the bristles are made from.

14. Organic cotton flannels – such as your grandmother used – are making a comeback. Strangely comforting and they just get thrown in the wash with the towels.

Also children love them. We don't need nylon 'scrunchie body puffs'.

15. Women's sanitary protection. Sanitary towels with plastic backing, plastic tampon applicators and even tampons without applicators contain plastic. These products clog up our sewers, rivers and seas and are found regularly on beaches by beach cleaners. There is a huge range of alternatives available – again just put 'alternative sanitary protection' into Ecosia and find what works for you. While I almost always advocate avoiding plastic, if a moon cup of some kind works for you it can save the planet from oodles of plastic that gets thrown away, and save you a fortune, so they are well worth purchasing.

16. Wet wipes of any kind. They contain plastic and are now a major pollutant of our rivers, because people flush them down the toilet. If you care about the planet, just don't buy them in any form. If you are really concerned about being 100 per cent clean, do what Arab households do. (Put 'bidet shower' into Wikipedia.) These can be fitted very cheaply and won't pollute our sewers. Or if you have money in the bank, be French and have a bidet fitted.

17. Washing powder. Buy an eco-friendly powder in an old-fashioned cardboard box. Use way less of it and do your washing on 30°C. Fabric softeners or anything at all that comes in plastic, including 'eco eggs' and various other bits of plastic nonsense – no thanks. Or you can stop washing any chemicals at all into our oceans by using soap nuts. These are a kind of nut grown sustainably in the Himalayan mountains and they contain saponin which is a natural detergent. I got mine from Green

Frog Botanic. My clothes are clean, they don't strip the colour and they have saved me money because they last for months.

18. Cleaning products. White vinegar (bought in a glass bottle) and bicarbonate of soda (bought in a packet) will do almost all your household cleaning. A damp cloth made from old pieces of cotton sheets or similar will do the dusting just fine. Have fun seeing just how little plastic you can buy – and how many chemicals you will avoid at the same time – and how much money you will save.

19. Washing up pads. I've had difficulty finding something that works as well as those plastic-based sponges (which are not, of course, made of sponge but of plastic) with one scratchy side and one smooth side. Loofah pads work wonderfully though. And loofah is a plant so when they are past their best you can put them in the home compost bin.

20. Straws. If you have a child under the age of seven then I'd say the occasional use of a paper straw is OK and fun at a birthday party. If you have certain disabilities, a straw is essential. For the rest of us adults who have been weaned – I think we can drink out of a glass without using a straw. You have to remember this every time you are out: ask – very clearly – when you order a drink, 'A juice with no straw please.' Or 'A screaming orgasm cocktail (whatever you love best) with no straw please.' Once they serve you the drink and the straw arrives, it's too late. If you say, 'I'm sorry – I didn't want the straw,' they will just throw it away. Every single time you order a cold drink you need to remind people not to give you a straw.

21. Body butters, etc. Just don't buy them in plastic pots. I'm lucky here as I have a friend who makes them. I give her glass jars and she fills them with her homemade body butter and returns them to me. When items need replacing, hunt around to find non-plastic alternatives. Slowly and imperceptibly your home will start to look and feel more – well, more real. You don't have to take my word for it. You'll experience it.
22. Food containers. If you need food containers for your house, please don't buy any more Tupperware. But stainless steel 'Elephant boxes' – they are just gorgeous. Take one with you if you are going out for dinner at a restaurant where you'd normally ask for a doggy bag to take home any left-over food. My local Indian is so good I always bring leftovers home but I don't want their plastic containers in my house.
23. Dry-cleaning. If you have clothes that need dry-cleaning, remove the hanger and the plastic cover before you leave the dry cleaner's and leave them there. Carry the clothes home over your arm. As well as using less plastic, you'll save your dry cleaner money, you'll help release the chemicals that are used to clean the clothes into the air on the way home instead of into your bedroom, and you'll save yourself a walk later to return the hangers. If you need to bring it all home (if it's always raining when you go to the dry-cleaner's), be sure to return the plastic covers as well as the hangers.
24. Lighters. Choose old-fashioned matches in boxes with designs that please you. Not only are plastic cigarette lighters one of the most common items found littering our beaches, but also, if you're a smoker, inhaling the butane gas which the lighter burns off when lit adds to

the collection of toxins you're inhaling. Plus matches are just nicer.

25. Toilet air fresheners. Also on the subject of matches: when you find a box, small or large with a label that pleases you, put some in the loo with an old ashtray that needs a job. If you strike a match it magically gets rid of any smell and there is no need to buy plastic aerosols with chemical perfumes in them.

26. Printing inks. Most of us print less and less. Hoorah. But when you do print (I'm assuming you buy recycled paper), set your printer to print on both sides of the paper as this will a) save paper and b) mean that you'll save money as it will be longer until you're forced to buy more ink in those plastic containers.

27. Glitter. I'm not suggesting – for those who love sparkles – that we need to live in a glitter-free world. But conventional glitter – made from a mixture of plastic and polyethylene terephthalate (PET) – washed down the drain from our basins into the rivers? Poor fish. These tiny specks pollute our waterways and our seas. Biodegradable glitter is made of eucalyptus tree extract and aluminium in some clever way. So, if you love wearing glitter, you can still do so … just make sure that you have the right kind of sparkle.

28. 'Paper' cups with plastic lids for take-out drinks. Can you imagine how our parents managed when they went to meet friends? They sat and talked in coffee shops and tea rooms, with drinks served in elegant crockery. And then the establishments did the washing up. I recommend doing your research with this one. Find a place to meet your friends that uses china cups and reward them with your patronage.

You may notice that in many cafés you now have to have a plastic lid on your cup even to carry the drink across the room – in case you spill it on yourself. 'Health and safety' (again). And, as I'm sure you know, those paper cups are mostly lined with plastic and like their lids are not biodegradable. If you absolutely have to carry your drink down the street (and I admit it's useful sometimes if you're in a hurry), then buy yourself a permanent small flask. (Your partner, grandparents, aunts, uncles or a friend will probably have one in the back of their kitchen cupboard that they rarely use.) They are fantastic and keep your tea or coffee hot all day. Most good cafés give you a discount if you have one. Or be more radical still and make your own coffee or tea before you leave the house and save yourself a small fortune. Just one take-away drink a day at around £2.50 x 365 days? I can do that much maths. It's £912.00 a year.

29. Clothes, shoes, handbags. In brief, let's avoid synthetic fabrics. More on joyful alternatives later. For now, if you absolutely have to shop, choose organic cotton.
30. Plastic pencils. Just no. Or highlighters. Coloured pencils do the job and they really are objects of joy. Just don't replace any of your stationery with plastic. I bet you have enough already anyway.
31. Plant pots. Garden centres are really taking a long time to catch up. There is a wonderful nursery called the Hairy Pot Plant Company in the UK that make their pots from the hair of coconut husks. But apart from them most garden centres still sell plants in plastic (with plastic plant labels too). The least we can do is take them back. My local centre has a big box for recycling them so that people who want plant pots can take them

home. So this is a start at least – while we ask for plants in biodegradable pots.
32. Dog poo bags. Not all 'biodegradable' dog poo bags are good news. Some of them just call themselves that because they break down faster than others into smaller pieces of plastic. I pick up unused dog poo bags almost every day that have been dropped in my local park and present them, as gifts, to surprised dog owners as I explain that I can't use them because I don't have a dog. If you want to take care of the planet, make sure your dog poo bags are plant-based – made of corn starch – then they really would break down and be harmless if left on the ground.
33. Blinds. If you are in the luxurious position of decorating your space, choose thick curtains over blinds where possible. Curtains are the old-fashioned way of insulating your home as the cold air collects between the window and the curtains, keeping you warmer. Blinds are often made of plastic. If you're a creative and clever with a sewing machine, many charity shops are full of amazing vintage curtains.
34. Coffee and tea. Nespresso coffee machines (owned by Nestlé) are designed so that you have to buy their significantly more expensive coffee – in pods. It's easier and cheaper if you just buy beans or ground coffee. They say that the pods should be recycled by being returned to them. If they're disposed of correctly, every single cup produces aluminium waste. The L'Or Espresso capsule, designed as an alternative to the Nespresso pods, are not aluminium – they are made of plastic. What insanity is all this? I have friends who use an old-fashioned French press. It will last a lifetime.

Your kitchen will smell of coffee, you'll save a fortune and the grounds go in the compost. Tea bags? Yes, so many of them contain plastic. Another madness. So, don't put tea bags in the compost. Do what your granny did. Have a real tea caddy with real tea in. Use a beautiful tea pot you love. And then you can do the 'mindful tea drinking' thing. And celebrate the wonder of a really great-tasting cup of tea in the most joyful way: by pouring it.

35. Plastic shower curtains. You have to be kidding. Just don't buy them. Everyone who has them seems to have a nightmare keeping them clean. They contain chemicals. And apart from anything else they are hideous. Save up and buy glass or you can even find them made of hemp. But plastic? Really – no.

36. Plastic laundry baskets and storage containers. My local charity shops always seem to be full of hand-made wicker baskets and picnic baskets of all shapes and sizes. They are SO much more beautiful.

TECHNOLOGY 1
Our Digital Footprint: The Illusion of a Cloud That I Recall

I suppose it shouldn't surprise me any more how many of the actions that we can take to help the environment also benefit us personally. As we live more simply, everything gets better. There is more time. There is more space.

A friend has just visited me whose digital life has become impossible. She has 50,000 emails in her in-box. I wish this was an exaggeration. This friend is highly successful – a professor – but she has never learned the necessity of hitting the 'delete' button on a daily basis. She is not alone. According to the Good Planet Foundation, the average US citizen has over 500 unread emails. Statistics are not available for the read ones. You can understand how this many unattended tasks would depress a human. It would make you not want to open your computer and deal with anything. Which of course makes the number of unattended demands go up even more. Digital overload, like a house with 50,000 too many objects in it, makes a person feel constantly overwhelmed. Many people feel that they can never meet all the demands that are being made of them. It's an exhausting way to live and more and more of us are existing like this.

OK, so lots of people don't do their digital housework, you may be thinking – but is there any real-world environmental impact? Yes, there is. By deleting what we don't need, unsubscribing, not hitting 'reply all' and simply by sending fewer emails, we are, once again, taking a direct action to lower CO_2 emissions. YouTube, WhatsApp, Twitter, Facebook, Instagram, TikTok, Chat GPT? Yes – all that too. But let's talk about email first.

Here are a few facts to inspire you. I've read that 28,397 tons of CO_2 is released into the atmosphere every day because of unwanted emails. A study was done (how on earth do they find out these things?) which revealed that in the UK alone 64 million unnecessary emails are sent each day. This includes newsletters from organizations that we have never signed up for or long ago stopped being interested in. You can't even buy toilet roll online these days without being put on a mailing list and being emailed forever asking you if you are ready to buy more toilet roll. You can't sign a petition in support of anything without being informed, for the rest of your life, how that organization is faring. It's enough to drive us crazy. And of course it does drive us crazy. If you don't take a second to hit 'unsubscribe' they will pursue you forever. Even after your death, they will still be notifying you of some special deal that they'd like you to take advantage of.

The connection between the deterioration of what remains of our sanity and the CO_2 in the atmosphere may not be immediately apparent. So I will do my best to explain the bare minimum of what we need to understand.

I hate to spell this out – to destroy the comforting illusion that has been created in our minds – but friends, there is no 'cloud'. There is no pretty, fluffy place up in the sky that our photos, messages and entire digital lives can be stored safely without impacting anyone or anything. It was so clever to call it 'the cloud', wasn't it? 'I'll save it to the cloud.' There can't be any harm in that, can there? Sigh.

The cloud – our storage facility – is actually mass data centres. In 2023 there were 10,978 of them worldwide. They are mostly huge concrete buildings. The average date centre is 100,000 square metres. There is one in Newport, Wales that is 2 million square feet. They contain seemingly endless rows of computers storing, well, everything. If you've never seen one, just put 'data centres' into Ecosia and hit 'images'. That fun video that you made of your cat? It's stored in one of these – somewhere.

There are more than a few problems with these huge centres. Poor America has 5,381 of them – the UK suffers less with only 514, but they are building new ones every day. Because our information has to be available for us 24/7 (after all, we may want to watch that cat video at 3 in the morning) the computers have to be constantly running. As a result they become very hot. So as well as using huge amounts of power to run these massive centres, more power has to be used to cool them down. Data centres in the USA use more than 90 billion kilowatt hours of electricity yearly. A single data centre can use more power than a medium-sized town. Someone has measured it. One data centre can consume the equivalent electricity of 50,000 homes.

And (please excuse me as I have to write this in caps as I'm still absorbing this information myself) THE CLOUD NOW HAS A GREATER CARBON FOOTPRINT

THAN THE AIRLINE INDUSTRY. Friends, the situation is very serious.

Then there is the impact on the communities of these places. If you put 'data centre' into Wikipedia you can read the words of human beings living by one of these centres. They describe the noise of the cooling fans as, 'A constant high pitched whirring noise that continues 24/7 ... like being on the tarmac with a plane running constantly, except that the plane keeps idling and never leaves.' The noise must be unbearable but of course they can't leave as their properties have lost their value. No one wants to live anywhere near a data centre.

Some of these horrendous buildings use water for their cooling systems. In Utah local residents are suffering water shortages and seeing their rivers drying up while the local data centre needs 7 million gallons of water daily to operate. You could be forgiven for thinking I must have these facts wrong. I wouldn't blame you.

Then of course there is the fossil fuel industry. You can often find it if you follow the money. Greenpeace has exposed many links between Big Tech and the fossil fuel industry. While some of the big players, Google and Facebook for example, are aiming at been carbon free by 2030, many of the data centres have no such aims. They run on electricity and when the electricity isn't renewable the financial benefits go straight to the fossil fuel companies. Huge growth in the need for data storage infrastructure is expected in the next decade. Some estimates say it will triple. As you read, Google is building a £790 million data centre in Hertfordshire. When the local residents bought their homes, they looked out on green fields. Yes, it's grim.

But – we are focusing on the difference that we can make. We will not be grimmed. As I said, it turns out that what we can do to help the environment will help us too. We can create digital space and mental space. And – breathe out ...

According to the Good Planet Foundation, if we take an evening and delete 500 emails, assuming that we don't need them and they are unimportant, this will remove approximately 175 grams of CO_2 from data storage. Of course you can argue that this won't make a difference to anything. Just as people say that taking one less plastic bag won't make a difference, one more person riding a bike instead of a car, one more person refusing to buy plastic won't impact the plastic industry. Yada, yada. I ignore all those complaints and keep going. All we can each do is all we can each do.

Assuming that each email contributes 0.3 grams of CO_2 to the footprint of data centres, I've read that if everyone removed just ten emails, it would remove 1,725,000 Gigabites of data stored in data centres. Of course I like to be radical – so I'd go further than removing ten. I'd like everyone to make a serious commitment to regularly and vigilantly removing ALL their unwanted emails. Everything that we no longer need, or that we never wanted in the first place.

Unsubscribing is an activity that brings enormous relief. All those newsletters that you may intend to read but don't. So much rubbish. A friend has a great system. She told me: 'I do it in batches. Whenever I unsubscribe from anything I make a note of them. If they go on emailing me after I've unsubscribed, I report them as spam.' Don't mess with this woman.

Sometimes it requires genuine persistence. Since I've been researching this I've been unsubscribing myself. I unsubscribed from Linkedin. Like most people, I'm there. It may have slipped my mind but I certainly don't recall asking them to send me emails. So I hit 'Unsubscribe'. The next day two more emails arrived from them. I was indignant until a friend pointed out that they have many different lists. They are emailing me 'career opportunities' that I haven't asked for. Then they are notifying me if anyone I know has posted on anything. Heaven knows what other list they have put me on. So I've had to unsubscribe five times (so far) and that's just one organization.

Another action that I recommend is examining very carefully what you may be storing on the cloud. If you have your photos backed up to the cloud, then yes, of course it's a good way to safeguard them, if the only other place they are is on your computer. But a few thoughts here. Please take the occasional evening and go through your photos too with your finger on the delete button. I think of it like this: if you were killed crossing the road tomorrow, no one is going to want to go through 20,000 photos – assuming that your loved ones even have your passwords. If they don't, those photos (as far as I know) stay on these servers for all eternity. Is there any plan to delete them after 100 years? Not as far as I know.

But even if your loved ones do have your passwords – don't you love them enough to NOT leave them 20,000 photos? You'd be leaving them a job that it would have been kinder to have done yourself. Just the same as decluttering your home. It would be an act of love on your part to go through them and just pick out the gems. Delete the ones where people have their eyes closed, that are out of focus, or where the leg of a stranger masks your auntie's smile. But, you may well ask, 'How about you, Ms "How to Practice Without Preaching"? How many do you have?' OK, I'll be totally honest. I currently have 7,017 photos. But they will go down. They will go right down. I'd love to leave 100. Maybe

500 as an absolute maximum. How many photos would you like of the life of your late grandmother? I'd love to have 100. Choose a figure that suits you but I'd be surprised if it's not considerably lower than the number you currently have. Imagine if we each left future generations 100 of our carefully hand-selected and much-loved photos? Now that selection would be worth backing up to the cloud.

Then there is the question of videos. I currently have 371 short videos. If you want to make deletions for the sake of your computer, your mental health, the sanity of those living by data centres, and the CO_2 in the atmosphere, it's best to start with videos. They use far more data than still images of course. In terms of CO_2 impact the worst offenders are YouTube and Facebook. Think of it – all those people endlessly uploading content while data centres are built on fields. Yes – it's a horror movie. So if you are one of those people who, years ago, made a fortune teaching people how to do make-up or who was paid well to give your opinion about books – well, you're probably now not so happy about what's up there … they have a button for people who have posted to YouTube called 'Delete forever'. And there is certainly a lot on YouTube that needs deleting.

Facebook is a difficult one. On the one hand you may want to ensure that loved ones have your password and be clear that you'd like your account deleted at your death. On the other hand some people like to put a black background to indicate that the person has died and leave the account as a kind of tribute page where the bereaved can talk to each other or, more weirdly perhaps, leave messages for the deceased. 'Just to let you know that I'm thinking of you today on your birthday.' Really? On Facebook? Still, other accounts don't even have the black background – they just stay there in limbo forever, with the last post taking on eternal significance. So the poor dying person ends up wondering whether they should use Facebook to say goodbye or not. And whether they want to be remembered for all time by last words written on social media.

I would imagine that most environmentalists, which includes you of course as you're holding this book, wouldn't dream of asking to have their name carved on a piece of stone to lie in a graveyard for centuries. The vast majority of the funerals that I've been to have been cremations; or, in one lovely exception, a friend was buried in a burial wood, but no stone was laid. She was returned to nature. She didn't want to leave anything. Considering herself a spiritual person, she would have believed that she has passed into the spiritual realm. All her possessions are distributed and gone. Her digital footprint, however, is still taking up space in a data centre somewhere. You get my point.

But back to the present and our point of power. The actions we can take. The 'small' difference that we can each make. Last night I deleted 100 emails. That's 30 grams of CO_2. I unsubscribed from 15 mailing lists. I checked what I had stored on the cloud and deleted 22 documents I no longer need. It took a little time but I removed 532 photos and 58 videos from my collection.

It's not necessary to imagine that this makes a small difference. We know it does. And I feel better too. Lighter, somehow.

ENERGY 1

Smart Meters, Candlelight and Sleeping Children

The lights have gone out in my house. Just my house. Not my neighbour's house. Before you ask, no, it's not the fuse box. And I know what you're thinking but, yes, I've paid the bill. The lights went off at 8pm last night and with them all the power to everything. Of course. Shock horror, even the Wi-Fi didn't work nor could I charge my phone. It did feel a bit like the end of the world.

A friend advised that I call my energy supplier and tell them it's an emergency. But I'm not elderly or on a life-support machine; I'm not going to die because of a night without electricity. So I sent messages to the other people in the house before they returned home for the night (sundry guests and visitors), explaining that there was no power. My South African flatmate was very relaxed. She just said, 'It happens in South Africa all the time.' A visitor from Texas was too jet-lagged to care and a third visitor from Scotland was polite enough to comment that the house looked pretty by candlelight.

SMART METERS, CANDLELIGHT AND SLEEPING CHILDREN

At 9.30pm everyone went to bed. It was cold. This story has two completely different points. I'll get to them ... bear with me.

This morning I rang my green energy supplier (I'm with Good Energy, which uses electricity from 100 per cent renewable sources) and told them I had no power and they promised to send someone in two to four hours. But, of course, like all suppliers, they use contractors as fixers. At about 4pm a man arrived and said that it was my meter that was the problem. I needed a new 'smart meter' fitted.

'But I don't want a smart meter.'

Many people say that smart meters pulse a microwave-like radiation through the air and claim they have a wide variety of negative impacts. The evidence is purely anecdotal, and government guidelines assure us that smart meters are safe, but with 23 Wi-Fi networks already going through my home and a mysterious mild tinnitus that doesn't trouble me in friends' houses, I'm not in the mood to take any risks.

Repetition appears necessary, 'I don't want a smart meter. Thank you.'

'I don't have any other kinds of meters.'

'Really?'

He phones his boss and is advised to leave for his next call. But he's not unfriendly. 'Tell them that you want a traditional credit meter. I have to admit I agree with you. Don't like the smart meters myself but you didn't hear that from me,' he says over his shoulder as he leaves.

I phone Good Energy. They apologize that he has left me with no power.

'You are my green-energy provider – haven't you heard the environmental warnings about smart meters?'

'We have. We're not recommending them, we are just trialling them with some customers.'

'I don't want one.'

'You can put it in "dumb mode" so it doesn't transmit.'

A smart meter in 'dumb mode' – sheesh, they are really laying on the pressure, aren't they? 'No, sorry, I don't want one.'

'Our suppliers say they are having trouble sourcing the traditional meters.'

'Is that so? Really? So what happens if I dig my heels in and sit in the dark, literally, and wait. How long would it take before they could source one?'

'I'll speak to my supervisor.'

She calls me back.

'Can I have a traditional meter then?'

'Yes, you can. We'll source one for you and have your power back on today.'

It turns out that the rules require them to go to 'all reasonable means' to persuade me to have a smart meter. So to avoid one you have to say, 'No, thank you.' Several times.

So they send another engineer. Who tells me that there is nothing wrong with my meter.

'Thing about these old meters. They don't tend to go wrong as there is nothing in them to go wrong.'

It appears to me that they have told me the source of the problem is the meter in order to fit a smart meter. Trust no one it seems.

He found the source of the problem, which was nothing to do with the meter, restored my power, shook my hand and vanished into the night.

So that's my first point – be persistent and determined.

And here is the second – the joyfully magical part. We had enjoyed being without power so much that when it got dark

SMART METERS, CANDLELIGHT AND SLEEPING CHILDREN

we didn't switch the lights on. Because we left the electricity off, the internet wasn't connected so as it got dark no one turned on screens of any kind. We found all the woollen clothes that we'd snuggled up in the night before and put them on. We lit the perfumed candles. We liked the house looking and smelling so mystical.

My Scottish visitor plays the guitar. And his voice, pretty standard when he speaks, becomes velvety when he sings. Much to the surprise and delight of my house guests, he sings, from memory, an entire repertoire of traditional Gaelic songs. We listened so happily, cosy under our blankets.

Then I remember that my Texan visitor is quite an advanced t'ai chi practitioner. I ask him if he wouldn't mind showing us the entire sequence. He's happy to. So, for the pleasure of the singing guitar player, my flatmate and me, he stands up, takes up the opening pose and then moves his way through the whole t'ai chi form while the candles make his shadows dance around him.

Then my South African flatmate announces that she can dance with fire poi. These are weights which are lit and then twirled around the head in circles and beautiful patterns. By night they are magical. So we stand outside, joined by a few neighbours, while she twirls flames around herself – so exciting and beautiful all at once.

A neighbour who thought that we still had no power brings us all hot drinks so we invite him in while I consider what I could contribute. Eventually I settled on reading some favourite poems. I read 'Like This' by Rumi, 'Somewhere in the State of Colorado' by Simon Armitage, 'At Lunchtime' by Roger McGough and I even do something I've never done before and read a poem I've written myself. Somehow by candlelight everything feels less intimidating. We are all half hidden after all and it's more fleeting. More sensual. More – well, valuable somehow. Because I suppose a candle burns out and everyone knows that. Just as our lives burn out.

We live most of the time as if energy, power, electricity and our lives will go on forever. But maybe through this incredibly beautiful miracle of 24/7 power we have all been robbed too. Robbed of the calm of the darkness. Of the beautiful black night. Of the long evenings. Of having to cuddle each other to keep warm.

And, I may be getting carried away here, but so many people suffer from insomnia. There is no 'calming down' time. There is no evening before the night. An ex of mine had two sons and he couldn't get them to go to bed in the evenings. They were always hyper. There are so many parents that have trouble getting their children to go to sleep.

Of course, their screens were never off. I remember he asked me for suggestions (foolish man) and so, slowly but surely, I started to calm down the atmosphere around them. We painted a wall light green, which is known to be calming. We hid some particularly noisy plastic toys. I tidied up, so the environment was less hectic and stressful. He took the phones away, not as a punishment but as part of our games with darkness. I turned down all the lights and lit candles. I played sleepy, calming music very low in the evenings. I burned lavender oil. My boyfriend played his guitar. We closed curtains. I put hot water bottles in their beds so they became cosy places to be. We gave them Calvin and Hobbes comics to read. They were allowed to read by candlelight. (They were old enough to do so safely.) And did all this calm them down? Mostly it did.

They went to sleep because of what we had created: an 'evening'. I'm not sure if the eyes tire more easily reading by candlelight or whether it was the soothing black of the darkness itself, but the spell worked and we loved our dark evenings.

The Asháninka tribe in the Amazonian rainforest live as they have always lived – with no artificial light. When I visited them, I didn't even see candles of any kind. So, at night when it got dark everyone went to sleep. And in the morning at first light everyone woke up. You never heard a child cry saying that it didn't want to go to sleep. Such an idea was unheard of. And at dawn, children were out with the adults gathering wood to make a fire to make breakfast. Of course they got up when it was light. It was only their Western visitors who, inexplicably to the Asháninka, would sleep after sunrise.

So what I'm suggesting is that you may like to play with this idea and find a version that works for you and your family. Maybe you could have occasional 'power cuts' and let your children experience life without electricity and gas. Maybe if you suffer from insomnia, you could experiment with getting up at 5am and going to sleep at 9pm in the winter when it's dark.

We are always being encouraged to explore new forms of energy saving, to be more aware of our energy use. But – as ever – I'm suggesting something far more radical and in a magical way far more joyful.

Just see how much you can switch off all together. Let day be day and night be night. Go outside and run in the morning when the sun shines. At night, when it's dark, go to bed. Preferably with someone you love. And keep warm under the covers of a cold room. Challenge your habits. Play these free, easy games.

ENERGY 2
Changing Light Bulbs on the *Titanic*

I looked up 'How many environmentalists does it take to change a lightbulb?' And the answer I found? 'None. If the bulb is blown then that's the way nature intended it.' I think that's meant to be an attack on tree-hugging hippies, so I must be a special kind of hippy because I like that answer. It's profoundly pleasing somehow. As you know, I love the dark.

But meanwhile, if you do want to light your property with electricity, here is a way that you can save 95 per cent of what you are spending on lighting and help the planet at the same time.

I have spent most of the last day two days reading about lightbulbs. Seriously, I have. I know I need to get out more. But I do find discovering new solutions genuinely exciting. Here's the brief …

First of all we had traditional lightbulbs. You know, the old 60-watt or 100-watt bulbs. All we had to remember when we went shopping was whether it was a bayonet or a screw-in. They blew frequently, but they gave a nice light and were cheap so no one cared or thought much more about them. These are called incandescent bulbs. They were supposed to be phased out across the European Union (EU) by 2012 but somehow they are still being sold. They leak 90 per cent of their energy as heat and they blow regularly.

Then along came those twirly energy-saving light bulbs and everyone who considered themselves even vaguely hippy slowly but surely replaced the old bulbs when they blew with the more expensive energy-saving ones. But lots of people hated them. They took time to warm up, they gave out a strange light, some people reported headaches, there were unverified rumours that they emit electromagnetic radiation and we learned that they contain mercury, which is toxic.

So then when the shitty twirly tubes broke or died, many people, for want of better information, just went back to incandescent bulbs which are inexplicably still available.

But then came a better solution. Enter the new LED bulbs or light-emitting diodes. Suddenly lighting became a very interesting subject. These bulbs have all sorts of astounding benefits. First of all, when you buy them you need to know about not only the brightness – the lumens – but also the colour temperature – the kelvins. In the morning we need a whiter, bluer light to help us wake up and concentrate and in the evening we need a warmer, yellower light to get us accustomed to the idea of sleep. And – get this – these bulbs can do both. They are called 'smart LED bulbs' (nothing to do with smart meters) and you can set them to match your circadian rhythms. This is especially good if you have trouble sleeping or have family or children who have trouble sleeping.

And that's not even the best part: these bulbs use 95 per cent less energy. Yes, really that much less energy. So, although they cost more (you'll be unsurprised to learn), they take a huge amount off your electricity bill and this means that, as households, we take substantially less energy from the national grid. The UK, to be fair, has made some progress in reducing its carbon footprint and apparently large numbers of people changing to LED lightbulbs (and also changing to electricity to heat their homes) has been responsible for one third of the reduction. In America, if each home changed just one incandescent lightbulb for an LED bulb, the country would save $600 million a year in energy costs.

There is one warning that goes alongside this – disposal. Red LEDs (those found in Christmas lights, car lights and brake lights) contain lead and white LEDs, while they contain less lead, also contain nickel, which is another hazardous waste. And so with both these and with the old energy-saving ones – if they break you should dispose of them carefully. Don't breathe the fumes and treat the insides as hazardous waste, not just something you can throw in your normal household non-recycling bin.

But having said that, LED light bulbs are designed to last for 25 years. Some of the better ones are even designed to last for a lifetime. So you shouldn't have to change them too often.

I'm hoping that I'll inspire you to look at every single bulb you have in the house. And as each one needs replacing – slowly but surely you will replace them with LED bulbs that are geared to your circadian rhythms. Just never buy the incandescent bulbs (or the halogen bulbs – they are way too expensive and waste loads of energy too because of the heat loss). Hopefully you'll be able to see how much money you are saving directly from your bills and you'll also be doing your part, as a household, to take less energy from the grid.

Then the answer to 'How many environmentalists does it take to change a lightbulb?' will be: 'I have LED bulbs. Come back in 25 years and I'll tell you.'

PLASTIC 4
Soap

A friend of mine enjoys making soap and selling it to her friends. Making your own soap means you can perfume it with the essential oils of your choice and avoid the chemicals (parabens and sulphates) that shop-bought soaps often contain. She doesn't bother with expensive packaging but just makes the soaps look beautiful using turmeric, paprika, cocoa powder or coffee to produce rich colours. I buy her soap, with delight. Once you have bought something homemade by a friend, you never want to buy any other way.

Hand sanitizer (usually sold in a small plastic bottle) is a new sign of our insanity as a human species. As if it's not bad enough that we buy water in plastic bottles and carry it around, and coffee in cups that we clutch like babies with dummies – now people buy hand sanitizer and carry it around with them. People carry the plastic bottles everywhere in their bags and are obsessively rubbing their hands with chemicals. Unless you work in a hospital or are living somewhere with no access to water, you simply don't need to buy these plastic bottles. And you know what?

SOAP

Although it kills germs it doesn't actually remove the dirt from your hands. Soap does both.

Then there is the subject of liquid soap. Some people I know are very attached to their liquid soap, in plastic bottles, with pumps. I have heard it claimed that pumps are 'more hygienic' and 'keep the bathroom cleaner'. I even – much to my amazement – heard someone I know say that although she could share a bar of soap with her husband, she didn't like the thought of leaving a bar in the bathroom because: 'I don't like the idea of it touching some visitor's armpit.' (Or presumably – even worse – their genitals.)

Friends, what has happened to us? This is a bar of soap we are talking about. How did we become so squeamish that some of us can't cope with the idea of sharing a bar of soap with strangers?

And yet, when I wrote my previous book *Sensation*, I heard plenty of stories about people who were prepared to have sex with strangers. One man tells me a story about meeting a woman for a date at a pizza restaurant and (according to him) she said, 'I don't really fancy pizza – can we just go and have sex?' And he thought, 'Why not?' So they skipped pizza and went back to his place and had sex. The telling of this story to friends on social media produced many and varied responses, including: 'Good for her', 'She could have been murdered,' and 'Why on earth did they skip the pizza?'

I find the story intimidating, as I could never imagine having sex with a stranger. But hey – I can imagine sharing a bar of soap with one. Am I excessively emancipated? In my own house, all the visitors to my bathroom use the same bar of soap. It's SOAP – it's self-cleaning! And yes, I do occasionally have to wipe the soap holder on which the soap is placed. But these are just the kind of radical actions I'm prepared to take on this environmental journey.

How did our grandparents manage to wash before plastic liquid-soap dispensers were invented? What uncivilized lives they led. In those days when there were fish in the sea.

FOOD 1

Holy Cows and Fish Fingers

The documentary *Cowspiracy* blows your mind. It just blew mine. It's in metaphorical splatters on the floor. This 'becoming an environmentalist' is hard. It was bad enough when I began to understand about the power of the petrochemical industry, but this has changed everything all over again.

The film points out that animal agriculture – that is, the breeding of animals as products, the agriculture that grows food to feed animals rather than humans, and the waste from those industries – causes more damage to the environment than pollution from all the cars, trucks, planes, boats, trains and ships put together. That sounds as if I've misunderstood, doesn't it? Apparently not.

According to this documentary – and they list all the sources of their information – 51 per cent of all greenhouse gasses are caused by animal agriculture. That got me reading, studying, researching and calling scientist friends. This figure is contested and some sources claim that it's nearer 15 per cent, but that is still as much as all the pollution from all forms of transport worldwide.

Why hasn't anyone told me this before? Why isn't this the very first thing you learn about when trying to figure out how you can help the environment? Why don't Greenpeace, Friends of the Earth, the RSPCA, WWF, the Nature Conservancy Council, the Rainforest Action Network, the Sierra Club, Surfers Against Sewage, the Royal Horticultural Society, Conservation International, the RSPB, Blue Planet Society and every single environmental group worldwide talk about this? They often have a dropdown menu on the home page of their website that reads: 'What can I do to help?' But it usually gives details on how you can take out a membership to the organization. First and foremost, it could say: 'The most valuable thing you can do is to explore having a plant-based diet.' That, as it turns out, is what it needs to say.

But you know already why they don't do that, don't you? It's not hard to guess: because if you are a meat eater it would piss you off. This subject creates visceral anger in many people. To ask people to change their behaviour counts as 'telling people what to do' or preaching. This would not be, for them, practising without preaching. And whole belief systems are based around our right to be free and live our lives as we choose: marry who we want to, work as we want to and eat what we want to – and heaven help anyone who tries to interfere in our personal choices. Or even to suggest that changing our habits would be a good idea for the earth. No membership organization in the world could afford to do that. Can you imagine what would happen if the National Trust in the UK announced that the best thing people can do to look after our countryside is give up their Sunday roast? People would probably resign their memberships so fast this would probably be the end of the National Trust. You wouldn't even see the skid marks as the 4x4s drove away. It's bad enough when we just ask people to stop killing foxes.

So, anyway, none of these charities mentions the Intergovernmental Panel on Climate Change report which says that people could help the planet by switching to plant-based diets. We are asked instead to do things that will not offend us. Recycle our plastic bags and join their organizations so that the professionals can deal with the problem of protecting the environment. And we can go on living just as we have always lived.

The *Cowspiracy* documentary claims that all the major international environmental charities know that this is the most important change that we can make in our lives but none of them have the courage to say it. Can this really be true? Can there be something this important which they are just not talking about? I had to go and study all the figures again.

Livestock and their by-products create millions of tons of carbon dioxide every year. Some sources say that land given over to livestock, and land given to grow food to feed livestock – rather than to feed humans – covers 45 per cent of the world's non-ice-covered land. One of the key reasons that people in the southern hemisphere are starving is because the land is used to grow food for cows not people. Hard to believe, isn't it? Children are starving – unnecessarily – while grain is fed to cows in intensive farming. Even if some of these figures are exaggeratedly high, there is no way to sugar the pill: the inconvenient truth is that one of the best things we can do to help the planet is to eat a plant-based diet. There is no avoiding this one. And because I'm not a membership-based organization, I can just be honest. Please be patient with me – remember I'm in shock too. Maybe we can't tell other people what to do but I have a duty to tell you the truth.

Of course, we can mess about with this one by saying, 'I'll go vegetarian on Mondays' and, yes, it would be a start. But then we'd still be part of the problem the other six days. Just as changing to vaping is, I suppose, better for you than smoking – but if you want to do what's best for your health, just facing the addiction and quitting nicotine for ever will do the job best. And so it is with our addiction to animal products. We have to be radical here. We're not messing about. Our environment is in trouble. Behaviour change is required. Not other people's behaviour – that's their job. Just our own.

Now, I know I promised you that we could become environmentalists joyfully – and yet I want to take your meat, your milk and your cheese away. I know that some of you hate me right now. This is hard because I don't want to alienate my readers. But what can I do? I have to tell you what I've learned and be honest with you. If you are still reading I know this may seem a step too far, especially if you've never eaten vegan food – or if you've eaten bad vegan food. But please stay with me: let me talk about this for a while.

So confession time: I've been a vegetarian for about 25 years. It was my daughter's fault. I was raised as a meat eater. My grandmother cooked for me as a child and gave me meat six days a week (due to the old Christian tradition, we never ate meat on Fridays), and I raised my daughter the same. I remember it clearly. It's many years ago. She is about six. She looks at her plate and says:

'What's this?'

'It's lamb, darling. Now be a good girl and eat it up.'

'A baby sheep?'

'That's right.'

'But why?' Her young brain works it out all too fast. 'Did it die?'

'Well, sort of.' I feel ashamed. I can't lie to my child. 'They killed it.'

'A little lamb?'

'Yes.'

'Why?'

'So that people could eat it. People eat lambs.'

'Why?'

What do you say to your child?

'People have eaten lambs for food for a long time.'

'Why?' I see her face crumple.

'I don't know, darling. It's not kind to the lambs. It's true. Don't be sad. They probably didn't hurt the lamb.' Oh dear – I'm not doing well here.

'When they made it dead?'

She bursts into tears. She refuses to eat the meat. She informs me, the following day, that she won't eat any animals ever. Strangely, I couldn't disagree with her but I found it very inconvenient. I couldn't even slip a Birds Eye fish finger by her. A week later, I tried.

'What's this?'

'It's a fish finger. Birds Eye. With breadcrumbs.'

'What's this inside?'

'It's – er – cod.'

'It's a fish? It used to be a fish? And they took it from the sea and killed it?'

I had to tell my child the truth. She refused to eat 'anything that used to have a face'. Then she progressed to, 'Did it used to move?' I never even tried to introduce shellfish.

So I had a vegetarian daughter but I went on eating meat and fish myself. I suppose it was more habit than anything. I had always eaten meat, chicken and fish and I liked the taste. I loved animals but there was a disconnect in my mind somewhere. And if I didn't think about it, the two weren't connected. I liked my Sunday roast as much as any meat eater.

Then one day I met a cow. I mean, really met a cow, not just going past one in a field while I was reading a book on the train. I was visiting a tiny children's farm and I got to have quality one-on-one time with a cow. Gandhi said that a cow is an expression of compassion. (Somewhat ironic, that we are killing an expression of compassion in intensive farms.) Anyway, I just noticed how lovely this cow was. And the next time someone offered me beef, even though I liked the taste, I just couldn't eat it. I imagined myself with a gun to the face of the cow I'd met, those big innocent eyes, that being I couldn't understand and yet knew she liked to be scratched behind the ears. I knew I couldn't pull the trigger. And it seemed unfair to me that someone else had to kill her for me. Talk about benefitting from the awful work that someone else had to do. No I couldn't be that person any more.

Once when I was in Morocco, I visited the Amazigh people in the Atlas Mountains and I asked a translator to ask a woman who had never left her village, and was about my age, when she was happiest. She replied in a second and I was delighted when the translator said,

'She says, "When I'm with my cows."'

I smiled, 'Not the men then?'

The translator laughed. 'She says – "No, definitely the cows."'

I felt envious that I had never had the friendship of a cow. I've known dogs and cats – but I've never been loved by a cow. I've never had a cow cross a field because I'm her favourite human. I've never raised or loved a cow. Maybe if I had I would also say that I'm happiest when I'm with cows. Alas for me, all I had done was let people kill them on my behalf and then eaten them with gravy. No wonder humans that live without animals are so lonely.

Pigs followed naturally. It's well known that they are more intelligent than dogs. They just have rubbish PR. Pigs are seen as dirty – as in dirty swine. A pig living in nature is no more dirty than any other animal and a pig swimming in the sea is as happy as you or I are when we're swimming in the sea. I couldn't kill a pig any more than I could kill a Labrador. And how many euphemisms have we created so we don't have to think about the fact that we are eating a pig? Ham, honey gammon, hog roast, bacon? People say 'I love bacon' without even thinking about the killing of a pig. Very soon after my decision to stop eating cows, I gave up eating pigs. My orthodox Jewish friends wondered what had taken me so long. My Muslim friends were also happy. Life is full of irony.

Do you know Tim Minchin's 'Peace Anthem for Palestine'? The lyrics are 'We don't eat pigs, You don't eat pigs. It's seems it's been that way forever. So if you don't eat pigs and we don't eat pigs, why not, not eat pigs together?' Check it out on YouTube – he's so worth it. But I digress.

Sheep? Lambs? I don't think I know a living soul who could kill a lamb. But I know plenty who eat them. I didn't have to see the terrible pictures. I was soon done with eating animals. Energetically, a lamb, a kitten and a puppy – what's the difference? They all love to play. I was changing; seeing connections differently. Looking at the wrapped meat in supermarkets with a strange new sense of profound sadness and loss.

Meanwhile, I lived next to a Hindu family from Gujarat in India. They had been vegetarian for as many generations as anyone could remember. The grandmother, in her eighties, did yoga every morning. The mother, my age, was a nurse and worked nights and the two sons both played football in the school teams. None of them seemed to lack energy. And, of course, their food tasted way better than anything I knew how to cook. I started to visit regularly. I learned the wonders of cumin and coriander, the mystical powers of asafoetida, and the fact that turmeric not only makes the rice a rich and wonderful colour but is a proven anti-inflammatory. You knew that, I'd guess, but I did not.

I pondered, for the first time, the phrase 'Let food be your medicine and medicine be your food' and realized I had no idea which foods were good for what. I had been eating animals that I knew had been pumped full of antibiotics and growth hormones – their meat full of preservatives. But somehow I'd just gone on without thinking about health. How would I make my food my medicine? My knowledge was limited to taking ginger and lemon for a cold and parsley contains Vitamin C … I knew relatively nothing about diet and nutrition. That's what years as a classic meat-and-two-veg eater had done to me.

Meanwhile, at my vegetarian friend Nirmala's house, all the wonders of the world were being served. And then I acquired a few more vegetarian friends – who also knew how to make food taste extraordinary. Once it happened like this: a visiting friend was trying on a dress. This involved her stripping down to bra and pants and, as she did so, I couldn't help noticing that there wasn't an ounce of fat or cellulite on her. She looked so fit and healthy. 'Goodness, sister-woman-person …' (I addressed her rather strangely, such was my admiration).

'What are you eating? Because whatever it is, I want to eat that.'

'I mostly eat salads,' she said. So, during her stay with me, I put her in charge of the salads. I had never eaten salads like these. The first one could have fed three people.

'How do you eat all that?' I asked.

'Slowly,' she explained. Have you ever taken half an hour to eat a salad? Her salads must have had 20 or 30 ingredients.

As I munched my way through nuts, seeds, fruit, vegetables, herbs, flowers and even random foraged leaves she'd picked in the garden, I began to realize that I'd probably been undernourished for years.

What I'm saying is that becoming a vegetarian was the beginning of my education in nutrition. My diet soon became much more varied and so much more enriched.

I was lucky with the daughter (I can only assume that she must have been a Jain in a previous lifetime) because whatever I put on her plate – as long as it had never had a face – she would be willing to try. I'd go anywhere they had fresh vegetables and buy anything I couldn't identify. The subject of the quality and variety of the fruit and vegetables was new too. I'd take home a red carrot, a turnip, an orange cauliflower, okra, multi-coloured pumpkins. I'd make a game of trying to find foods that she hadn't seen before. She ate everything.

How do you feel about cooking? Given half a chance I'm lazy with food. It's not that I don't enjoy it; it's just that I'd rather be doing almost anything else. I'd rather be writing to you. So, cooking doesn't come naturally. But I've realized that if I'm going to do this properly – radically – then one of the most important things I can do is embrace a plant-based diet. And to do so joyfully.

Dairy products too? Yes – it's time. Even though that's also inconvenient, I will choose to embrace a fully plant-based diet joyfully. And it won't be hard because – in the back of my mind – I know that intensive dairy farms are cruel. The practice of taking a calf from the mother at two days old (which I believe is standard) is something that would make me weep if I saw it with my own eyes. And, from those who have done so, I have heard that the sound a cow makes lowing for her calf is so heart-wrenching it's not something that you'd ever want to hear. Dairy cows that in nature live to 20 years old, are killed at 5 to 6 years old. They have been modified to produce up to 10 times more milk than they would naturally and, apparently, at least one third of them, at any one time, have mastitis. I'm glad I don't have to see this sort of suffering. I'd be shouting, 'Don't put that cow in that machine. Can't you see that her udders are infected?' I'm just glad I'm not a part of it. Any more.

It's about being in alignment – being congruous. Not doing anything, or as little as we possibly can, that causes suffering. As they say so powerfully in *Cowspiracy*, it would be impossible for me to become an environmentalist and not also become a vegan. With my food choices I get three votes a day and I want to make every single change to myself that I possibly can – because that is my point of power.

It's an excuse to make a lot of new friends, have a lot of parties and even buy some new cookbooks. And Instagram is the real home of the vegans, I've discovered. It's where vegans share glorious multi-coloured meal ideas every day – thousands of them. Vegans on Insta typically have many more followers than anyone else.

And it feels good to finally arrive at being vegan. It has been a long time coming. Vegan friends have been horrified as I've gone on buying dairy. And it's never felt right. After all, I'm not a baby cow.

This isn't a book about the joy of not adding to the cruelty in our intensive farming industry though – it is about us becoming environmentalists. So, as part of taking every single action that we can in order to save the planet, it may be helpful to know that according to the scientific author Joseph Poore and a group of researchers at Oxford University, adopting a vegan diet can reduce your carbon footprint by up to 73 per cent.

Joseph initially began researching sustainable meat and dairy production but couldn't avoid the information that he and his team of researchers uncovered. Five years later they concluded that avoiding the consumption of meat and dairy products completely produces a far greater environmental impact than attempting to purchase sustainable meat and dairy. He found that fresh-water fish farming and grass-fed beef also pose catastrophic environmental problems. After the first year of his project he stopped eating all animal products. Really, it's a no-brainer. And if you're doing research at Oxford University, using your brain is a vital requirement.

So here I am. That's it. Plant-based eating for me from now on. What a fresh, nutritious, colourful, organic joy that will be. Some sources say that there are now over 3.5 million vegans in the UK and the figure is rising every day. I won't be alone.

FOOD 2
How to Buy a Tomato – and Other Supply Chain Dilemmas

Having decided both for the love of animals and the love of the planet not to eat animals and for the love of the sea not to eat fish, the question remains: 'Who grows my food?' The question 'Who makes my clothes?' is well known in the fashion industry, but really learning about where our food comes from is not only a great tracking game for families to teach their children but, as it turns out, a very enriching adventure for the grown-ups too.

In my local area I have an abundance of supermarkets. As an experiment I went out a while ago to see what I could buy that isn't sold in plastic, is grown in the UK and is organic. It proved to be an exercise of limited visual appeal as I was posting photos of the results of my research on Instagram. The pictures were: Tesco – an empty basket; Lidl – an empty basket; Asda – spring onions; Sainsbury's – an empty basket; Iceland – an empty basket; a local independent Hoxton

Fruit and Veg – a very attractive wicker basket but empty nevertheless; Waitrose – spring onions. The only two shops locally that were able to give me enough UK-grown seasonal vegetables to live on were Bayley and Sage – an expensive chain which seems designed to cater for the super-wealthy – and Whole Foods – the very same store I started this book by promising to avoid. And now of course, Whole Foods is owned by Amazon.

Having committed, for the purpose of this book, to doing everything I can to make the most joyful environmental choices, I sigh wearily, learn how to make spring onion soup and complain to everyone about the situation regarding packaging and organic food.

One morning a box appears on my doorstep. It is labelled 'UK, seasonal, organic'. And there isn't any plastic in sight. Oh joy. I had tried box schemes ages ago when they were less efficient and organized. I'm concerned that this isn't a solution because it's more expensive than shop-bought food. 'Live life on the veg' is printed on the side of the box. It's a company called Riverford. I ring them up to see where it's come from. My daughter has sent it.

I phone her. 'But is this really an environmental solution?' I ask her. 'What about the road miles?'

'Well, it's the only way I could find to get you real organic food that is seasonal and doesn't come in plastic. I've ordered you a box every two weeks. It's for my own peace of mind. I need to know that you're not going to starve.'

'Why Riverford? Are they any good?'

'Reviews are good. Articles are good. The company is owned by those who work there – like John Lewis. That seems good.'

'Isn't it expensive?'

'It's certainly not more expensive than some of the shops you've been to recently. Anyway, you'll like the seasonal vegetables. And you'll have to learn how to cook them.'

It was true. They had sent some kohlrabi, nicknamed the 'alien vegetable' as it looks like an alien. I have no idea how to cook it. 'Now if you could just send a chef …'

'Mother …' the daughter assumes a scolding voice as if dealing with a recalcitrant child. 'Just put "kohlrabi recipes" into Ecosia and they will come up. Anyway, Riverford send recipe ideas with the food. Why don't you go and visit them if you really want to learn about the supply chain?'

Now, that's a very good idea. Then I could solve this 'where to shop' problem. I write to Guy Singh-Watson, the founder, and to my delight he agrees to an interview.

I travel down to south Devon to see where the vegetables I'll be eating have been grown and to learn a little about what it takes to grow organic fruit and vegetables.

My lovely Airbnb host in Totnes kindly drives me to the Riverford farm. I sit on a bench outside their kitchen restaurant surrounded by growing herbs and flowers. It's a sunny morning and I'm brought coffee. All feels right with the world. Then Guy arrives. He's a tall, good-looking man with rugged features. He looks like a farmer and dresses like one. He doesn't look happy this morning. Maybe he'd rather be out in the fields working on the land than doing interviews with strangers. His dog is friendly though and wags its tail at me.

So here is our interview. Me: smiley, happy, glad to be there. Guy: not so much. He tells me that he's in a grumpy mood this morning but reassures me that it's nothing to do with me. I sip my coffee, switch on my recorder and begin. I hope I've asked all the questions that you'd ask.

*

'I'm very happy to meet you, Guy, and your gorgeous spaniel.'

'Thank you.'

'To start with the basics, I understand, in the most simple way, that we need to buy organic food because we need to understand where our food comes from. If we don't buy organic then our farmers will go on spraying chemicals on the food and this harms the insects, the soil and us. But I wonder if you could say a little bit more to any reader who feels that there is no real difference between organic and non-organic.'

'I'm never going to claim that "organic" has all the answers or that it's the only answer. There are some arguments for chemicals in agriculture and organic isn't perfect. However, I do think it's the best show in town because it's the only mark, the only methodology that communicates, in a way that is trusted, what is happening on the farm. It gives customers the opportunity to exert their preferences and choices about the way that land is farmed. At the moment there is no other way. You can choose to buy free-range eggs but that's very limited and there is the Pasture-Fed Livestock Association which I think is a great thing.'

'The radical concept of cows eating grass?'

'Exactly. It's about getting ruminants to eat what their digestive systems are designed to eat – namely grass – rather than feeding them grain and soya. From an environmental point of view this is very important. That's another mark but it's not yet very well known.'

I have to ask him about the most frequently raised objection to organic food – the extra cost. I begin diplomatically, 'Personally I can't afford to run a car or buy new clothes but I feel I can't afford not to buy organic food as I see it as vital for my health and for the planet.'

I suppose I then become less diplomatic. 'But take, for example, the Airbnb host, where I'm staying locally. I asked her about Riverford and she says that she can't afford to buy from you.'

'I wonder how she has reached that conclusion. I would guess that she's never bought a vegetable box. She's never been to the farm shop.'

'She's been to Lidl and she's been to the farm shop.'

'Well, the farm shop isn't me. So her position is ignorant and prejudiced – like so many people's.'

That's a bit strong. He's sees me flinch. And goes on, 'I'm sorry, but it is.'

'If there are people around the country that are on a very low budget … do you have any thoughts on convincing them of the importance of spending a higher percentage of a limited income on buying organic food?'

He explodes like a firework. (Light the blue paper and retreat.) 'I've struggled with that one for 30 years and I'm fed up with it. You know that 0.7 per cent of GDP goes to farmers? If it was all organic it might be 0.8 per cent. 0.1 per cent of GDP would make the difference. The money spent on food goes to retailers and the food service sector. Ninety per cent of the money you spend doesn't go to farmers.'

'I didn't know this, no.'

'Well it's a shame that more people don't know it. Really! I'm sorry but I get asked this question over and over again.'

'It's just that I know that a lot of people ask about the cost.'

'I was asked it the other day at a talk I gave and most recently by the BBC's "Food Programme". It's extremely frustrating being written off as some sort of niche for the privileged and meanwhile the world goes to hell in a handcart.'

'Or goes "to hell in a hybrid" – as they say at Friends of the Earth.'

'Quite. And we are told that we can't afford to produce food without neonicotinoids and without nitrogenous fertilizers, and so on. It's just not as simple as that because people are paying for all those things in other ways. We are killing all the insects, for example.'

'I know. I'm one of your customers.'

'I'm not having a go at you, I'm having a go at the question. The woman who runs your B&B, for example. If she bought one of our vegetable boxes and bought the same from Lidl and the difference is more than £2 a week, I'd be astounded.'

'I'll tell her.'

'If you choose to eat organic fillet steak then it's bloody expensive. Even organic chickens are relatively expensive.'

'She's vegetarian.'

'Organic vegetables really aren't expensive. It's just an old prejudice. I wouldn't mind betting that her mobile phone and internet subscription are way more than the cost of the vegetable box. We all make choices. Some things which are much better quality and are produced in a responsible way are more expensive.'

'Yes.' Actually I love his passion for this subject. I'd be angry too if I were him.

'The only way that we can get away from this is if we start internalizing the costs conventional agriculture places on all of us in terms of pollution, loss of wildlife, loss of biodiversity and climate change. If we counted all this then organic food would be cheap. But I don't see much hope of that happening.'

'We're doing our best, Guy.' Next question from me: 'If someone is not a customer of yours, it's very confusing when you go to the supermarket and find organic food in plastic that has been flown in from who knows where.'

'Nothing is flown in from anywhere in Europe. It's shipped in or trucked in but not flown. It may be flown from other parts of the world but not from Europe.'

'OK.'

'So your question is?'

'Well, just that it's confusing as it all looks the same whoever it's from and it's so hard to know how best to choose.'

'It's a hard choice. There are so many things to consider – the plastics, how far the food has travelled and how much synthetic fertilizer has been used on the land. That's why I don't sell to supermarkets. And I made that decision 25 years ago. Also, organic food is more packaged.'

'Yes, why is that?'

'It's done for the convenience of the retailer. So it's not confused. That's the reality around plastic usage. Sometimes it's to preserve the product and prevent waste but mainly it's just for the retailer's convenience and to add to its appeal. It's for their own fucking convenience and because they think they'll sell more. Personally, I just can't pick that food up in a supermarket. Does it make any difference if it's organic? A little bit, I suppose.'

'I see. So you just don't like supermarkets? I can understand that if you're selling real food.'

'And another thing: I'm very sceptical about all these outlets that aren't certified organic but make claims like 'non sprayed' and 'environmentally grown' because I've heard it so many times and when you scratch the surface you find that they are using every bit as many chemicals.'

'So if a seller advertises their food as "naturally grown" you'd be sceptical?'

'I'd ask them what the hell they mean by that. Anyone that says, "I'm as good as organic" or "I'm better than organic but I can't afford to be certified" just makes me spit tacks. It's not that expensive. I think it starts at about £300 a year and I have a business with getting on for £1 million turnover, so to spend £700 being certified – I really don't think that's a lot. This coffee they have brought us is insipid.'

'Do you want to order a stronger coffee that you'll enjoy?'

'No.'

He's on a roll. I rather like the rage. As the comedian Mark Thomas says, 'If you're not angry you're not paying attention.'

'The trouble with the supermarkets is that I see them as just as much a part of the problem as conventional farming is. Possibly more so in the way they encourage us to view food.'

'It's very joyless. In plastic and often tasteless.'

'Exactly. I heard Stuart Rose talk last week, he's ex-Marks and Spencer. He was giving a talk before me at a conference. His mantra was 'the customer is king' and we have to do whatever we have to do to please the customer. Given that we have a government that is completely useless and, in the foreseeable timeframe, isn't going to do anything and given that this is also the mantra of retail with almost no exceptions, the only hope is this is not what people want. They've got it all wrong.'

'I think they have.'

'If you look at what people really want – which is real food and to come to a place like this and walk around and have a connection with the land and a sense of where their food comes from, and know that it hasn't had the hell sprayed out of it and isn't destroying the environment.'

'Yes …'

'And to buy food from people that they can identify with …'

'And trust.'

'As you say – and trust. That's what people want. And there is a spectrum between people who come here and know us and understand what we are doing and really enjoy the taste of seasonal vegetables, and what the supermarkets offer – which is completely anonymous food from fake farms. And it's over-packaged and over-travelled. And there is no fucking choice for 95 per cent of it. So the idea that people are exerting their influence and saving they world while shopping in supermarkets and using 'informed consumer choice' – it's just bullshit. There is no choice.'

'We need to buy our food differently. I do understand.'

'I do think we offer a genuine alternative and there are a few others like us. A precious few.'

'Which others do you recommend apart from yourselves?'

'There are lots of really good small box schemes: Growing Communities is fantastic, Abel and Cole do a very good job. I don't think they are as good as us but then I would say that, wouldn't I?'

'I think it's nice that you're prepared to recommend your competitors. Let's just stick to the ones that you recommend, shall we?'

'Regarding the ones that do recipe boxes … Mindful Chef does a reasonable job. But I'd say that all the small box schemes that have survived are pretty good now otherwise they wouldn't have survived. Better Food Company in Bristol does a great job. Acorn Grocery in Manchester is a fantastic business. And there are other ones and a lot of them are doing really well but they are still a tiny part of the overall picture. And the ones that are doing well want to stay independent and they don't have the benefit of venture capitalist backing – which some other companies that I'm more sceptical about do – and this makes it hard for the rest of us.'

'Are there any products that either you or some of your suppliers grow that you are particularly proud of and would like to tell us about?'

'Yes, the peppers that we grow on the farm in the South of France are fantastic.'

'Padrón?'

'They are a green pepper. One in ten is bloody hot and the rest have a delicious flavour. Our artichokes and our carrots and potatoes are fantastic. Across the range the standard is high. I can't say that everything is good. How you grow a really tasty vegetable doesn't always work out. The blood oranges that we get from Sicily are exceptional, the Seville oranges from a grower near Seville are really, really good and the mangos we get from Spain are excellent – although their season is very short.'

'I must eat all these really soon. So, just to be clear, you don't sell anything that is flown to the UK?'

'No. We don't sell anything that is air freighted and we don't sell anything that's come out of heated glass. We did a study with Exeter University that considered what the real insanities are if you are going to try to produce food in a more environmentally sensitive way. Our carbon footprint and climate change is the key metric. I know that plastic gets some people fired up but carbon dioxide emissions or the equivalent – methane – is what we should be focusing on and if you do that then it's a no to heated glass and air freight. Frost protection of orchards is another dreadful one (we haven't yet said that we won't buy from anyone who does this but I think we will). Then there's thermal sterilization of the soil: we are very close to banning suppliers who do this. We already say that if this is a routine part of how you grow, we won't buy from you. These are things that customers don't know about but they consume vast amounts of fossil fuels.'

'That's why I'm here. I'm learning at least.'

'We are currently working again with Exeter University to look at the environmental impact of the business and I hope that we will move further down. Ultimately, we won't sell anything that has a carbon footprint of more than so many grams of CO_2 per kilogram. We haven't formerly set that level but we will do and then we'll work to keep bringing that down. Then maybe try to offset the rest with agricultural practices from both us and our suppliers that are sequestering carbon. We're looking at that and I want to make sure it will be genuine before making that claim but it might involve planting a very substantial acreage of mixed trees. Traditionally, in this area, sheep grazing under a cider orchard is very successful but, due to the structure of the industry this has been virtually abandoned for no good reason.'

'If I were a sheep, I'd rather graze in an orchard than in a bare field.'

'You have an apple orchard of standard trees, with 10-metre spacing, big apple trees for cider and instead of spraying off or mowing you just have sheep grazing underneath as they have done for centuries. Once every farm around here would have had orchards like that. That was when everyone drank cider rather than beer. It's that sort of system that we need to get back to. We're also looking at a system for growing nuts.'

'Nuts?'

'Yes, well, we haven't got till 2050. We've maybe got ten years. No one really knows but we have to, very quickly, get down to zero carbon. Everyone has to do that. Every business, every individual. How are we going to do that? Our carbon footprint as a business is 10,000 tons of carbon a year. It's horrendous. Can we get that down to 5,000 tons or even less? And then offset the rest with agricultural practices that sequester carbon? It's a challenge.'

'Ten thousand tons a year?'

'Yes. That includes all our employers driving to work. Every single delivery, every bit of transport from the farm to the customer. Everything.'

So that is his work. Back to our work.

'What are some of the things that people can do, in relation to their food, to help the environment most? Or what pisses you off most? You're not a fan of supermarkets. What else?'

'I'd start with the absolute number one – and some people would disagree with me on this but the absolutely most important step is …' He paused playfully. I tried to guess what he was going to say next. 'To eat less meat and less dairy products.'

I decided against asking him why Riverford is still selling meat and dairy then. Instead I went on listening.

'There are arguments that grazing livestock can be a good thing for the environment and I accept those arguments up to a point, but we can't eat meat. There are just not enough acres. So less meat. I notice you ordered oat milk for your coffee.'

'Yes, I'm vegan.'

'I'm not. I'm not even vegetarian but to my mind the reasons for eating less meat are overwhelming. The idea that it's somehow better to eat chicken and pork – it's fucking well not. Perhaps beef and lamb have been unfairly demonized if you are going to eat meat, but basically …'

'Eat less meat.'

'Yes. Eat less meat and include dairy and eggs in that. Less animal protein.'

How interesting that he has this stance but still isn't vegan. Ah well. Each in their own time. For a non-vegan, there he is – still advocating this as the most important thing that we can do. He continues.

'Second essential: wherever possible try to buy food from people that you really trust.'

I trust this man. Because he calls a spade a spade and not an agricultural digging implement.

'Very good.'

'And the third most important change, environmentally speaking, which I know is an incredibly difficult ask for most people but seek it out where you can: if you have the opportunity to buy organic, look after your planet and your health by getting your bloody money out and paying for it.'

I laughed. 'And if you're buying fruit and veg through a box delivery scheme it may not be that much more anyway.'

'Ask difficult questions would be my fourth point. Wherever you buy your food. And don't be fobbed off with absurd answers. When your local supermarket implies the food comes from whatever their fake fucking farm is, go and ask at the till – and write them a letter – "Where is this farm? I'd really like to know about this farm. I'd like to go and visit it."'

'Brilliant.'

'And my fifth point is also about asking difficult questions and persevering till you get answers. When you go to a restaurant, especially if it's a favourite that you go to regularly, and they tell you that the food is locally grown, seasonal and organic … ask them about it. "What exactly is organic on this menu? What is seasonal?" And just really be a pain in the arse. Because we are living in this "post-truth" time when it seems that people can just lie. So if they lie – and you go on asking questions – are you the one who is being a pain in the arse? No. A lot of them, if they are not actually telling lies, are deceiving by implication. If someone is lying to you or not answering your questions – be a pain in the arse.'

'An essential skill for the environmentalist.'

'If the waitress comes back and tells you that the vegetables are all slow grown and the birds are exclusively hand-plucked within the M25 … ask for some detail.'

I'm laughing again. He's definitely grown on me.

'That's an actual quote from a famous London restaurant quite recently. Until there is some degree of honesty and transparency, there is no way that customer preferences go back down the value chain to reach the farmers. It's all just lost with bullshit in the supply chain.'

'I'm very interested in supply chains with everything we buy. I'm even going to ask if we can have more information on the Riverford website about the supply chains. It's not always included where everything comes from.'

'I think it's mostly there. I'll check. Sometimes at the beginning of the week you are selling oranges from Italy and at the end of the week they are coming from Spain. But we should certainly monitor what percentage of our product comes from abroad.'

'I've been having your UK seasonal box plus apples. But the apples – out of season – come from Argentina.'

'I don't suppose they are very good, are they?'

I've eaten apples with more flavour but those are in season and grown in the UK. So I keep quiet.

'We sometimes sell apples from New Zealand and even though they are shipped I'm not happy with how far they come. They also use frost protection in their orchards.'

'As an apple eating customer, it strikes me that I could grow up a little and accept that I'm spoiled and I don't have to be able to eat apples all year round. I could just enjoy other foods and wait for apples to come into season.'

'In a couple of weeks the first Discovery apples will be ready. But not supplying certain products at certain times of year has a very significant impact on us and on our staff and our franchises. We'd probably knock 40p off our average order value and that hurts us financially quite significantly.'

'Just by not supplying apples from New Zealand?'

'In the long run it may help us as we'd be leading the pack and we'd be seen to be standing for something.'

'I think it would be great if your website said, "The new apples aren't ripe for another two weeks so you're just going to have to wait. Or we're going to have to bring them all the way from New Zealand and there are so many other types of fruit."'

'This month the apricots are superb, the English plums are just coming into season, the stone fruit is at its absolute best. This is why when you get a Riverford box, we decide what goes in it. This is counter to conventional retail.'

'Well, you know more about food than we do.'

'Thank you for saying that and, yes, we do.'

'The customer isn't always right.'

'But asking the customer to forgo their power to choose is very counter-intuitive. It's all about "empowering the customer". And we disempower them. If you eat in the restaurant here we'll tell you what you're going to eat, when you're going to eat and who you're going to sit next to.

We can take a certain amount of risk, like not doing this bollocks fake discounting for new customers that always has a catch in the small print – because we don't have external shareholders to worry about. We can treat our customers with genuine respect.'

'A random question: can you tell me how to buy a tomato that I can be certain hasn't been grown in a hot-house? I'm confused as I've heard you say that Spanish tomatoes are often better than UK tomatoes?'

'Yes. Don't buy a UK-grown tomato from the end of October through to the end of May – probably the end of June. The ones that you'll buy in November and December will be completely tasteless. They have no sunshine and you can't grow a decent tomato without sunshine. You can heat the hell out of them and they will still ripen. You can maybe even put in supplementary lighting. So don't buy a UK tomato in January. If it's minus 1 outside, it has been heated to 20 per cent in a single-glazed hothouse.'

'But as consumers, how do we know which is which?'

'Common sense. Even in a supermarket it will say what the country of origin is.'

'So, are you saying don't buy UK tomatoes at all if they are out of season?'

'Yes, I am. Don't support hot-housed tomatoes so don't buy them outside June, July, August, September and October. And get them as a treat, sometimes, out of season, from Spain.'

'Where are the hot-houses that we want to avoid?'

'In the UK, and a huge amount are Dutch. And this also applies to peppers. They are even worse, environmentally, out of a hot-house, as are aubergines and to a lesser extent cucumbers. All the things that people have got used to eating 12 months of the year. To my mind farming is about harnessing the sun's energy to make food. Not artificially heating and lighting huge greenhouses. When we did the study with Exeter University it was ten times worse to buy a UK-grown pepper out of season than to buy a Spanish one and five times worse to buy a tomato.'

Farming is about harnessing the sun's energy to make food. I'm beginning to understand why he was grumpy before meeting me. He must have known in advance all the stupid questions I was going to ask.

'Another question about the sun: am I right in thinking that I may buy a product somewhere not from you and it will say "organic" on it and it still won't have seen the sun or been grown in soil?'

'No, I don't think so. The standards of the Soil Association are very strict. The food has to be grown in the soil. I'm not sure about lighting – you'll have to check that with the Soil Association.'

He stood up.

'Now is there anything specific you'd like to see on the farm?'

'Yes. I'd love to see all the lettuces. I eat them so I'd love to see them growing.'

'OK, that's quite close so we can do that.' Guy's springer spaniel springs up and bounds around excitedly.

'What a gorgeous dog?'

'He's a rescue dog.' You have to love a man who has a rescue dog.

We walk through the trees that surround their kitchen and out into the fields.

'How much land do you have here?'

'If you add it all together at Riverford – myself and my brother and my sister – in total we have about 1,500 acres. As I'm sure you know, my family sold the farm to the employees in a joint ownership scheme a year ago, so Riverford Organic Farmers has about 300 acres and includes that hill. A lot of what's left is great for the wildlife but not much for growing vegetables. In that wood up there we harvest an amazing amount of wild garlic and we manage the undergrowth to favour the garlic. In many ways it's one of the most productive areas of the farm.'

'How long have you been here?'

'On this farm since 1996.'

'And do you still enjoy it?'

'Yes. I realize I'm incredibly lucky to do something that I love and that I have, largely, such autonomy over my life and what I do each day – and to be able to make a difference. To have ideas about the way I'd like the world to be and to have the power and independence to be able to implement them.'

'Hooray.'

'I'm very aware of that. I do count my blessings. I do a lot of public speaking and I'm not normally as grumpy as I am today – it's nothing to do with you.'

Phew. 'That's a relief.'

'No, really. I see from when I do public speaking that the appetite people have for something better is huge … Something that reflects better on us as human beings and, I would say, more accurately. We are so much better than our institutions, our governments, our businesses and our supermarkets allow us to be.'

'I agree whole-heartedly, Guy. And there are more and more of us looking for alternatives.'

The land we walk through is beautiful. Unlike so many farms that are just mono-crop fields with the odd tree around the edge, here generous amounts of space are left wild. Flowers grow along the banks.

'Artie!' He calls his dog. 'You are not allowed to run through the lettuces.'

'Wow. These lettuces are beautiful.'

'These are Little Gems, these are a green Batavia, these are a Cos type but they are a red one. Here, taste this.'

'This tastes rich. Of life somehow. It's alive.'

'I was visiting a farm in Lincolnshire, a very nice farmer as I remember. Maybe I was trying to convince him to go organic – I don't remember why I was there – and I bent down, as I do, to taste his lettuces. I'll often pick off a leaf to sample the flavour. And he said, "Oh, I wouldn't do that If I were you. They were sprayed yesterday."'

'How awful.'

'It's one of my great pleasures, walking around the fields picking vegetables for my supper for friends and family. And to think that what you're growing, the very sustenance that you are creating is covered in nerve poisons and you can't let people eat it in the fields just seems tragic.'

'What's this one?'

'This is a green Batavia. They aren't my favourites. I always think they have a slightly muddy flavour. Or maybe I'm being unfair to them. We grow them at the beginning and the end of the season.'

'I could eat one of these a day.'

'Yes. I can't understand how people can survive with a bag of 100 grams of salad. My wife and I must eat 500 grams of salad at least once a day and sometimes twice.'

'I'd love to do that.'

'That field with the cows on is the one I'm thinking of planting up with nuts this winter.'

'These verges are pretty.'

'We work with Devon Wildlife Trust. They have to sow yellow rattle first as the wild-flowers thrive in poor fertility.'

'What's this growing here?'

'That's spinach.'

'Can we have a photo with these amazing lettuces growing behind us?'

Guy tolerates me taking a few selfies. Then we walk silently for a while just enjoying the sunshine and the green around us.

'This is a cardoon field. I love growing cardoons.'

'Looks as though it's related to thistles and artichokes.'

'It is, yes. You eat the midriff of the leaf and you can only eat it at certain times of the year. I have a lot of trouble persuading people to eat them. It's a standing joke around here: Guy and his cardoons. But this field will be sequestering a lot of carbon into the soil. And the bees love them. We sometimes sell the flowers.'

'Can I take some back to my landlady?'

'Of course.' He picks three huge cardoons by the stems.

'Can I give a short speech about how impressed I am with all of this, Guy? You've created work, created employment, fed people – and all without damaging the planet. This is all such a fantastic achievement.'

'I do feel very proud but also lucky to have had the opportunity to do my own thing and to have been supported by people who shared wanting to see that thing happen. Is that everything?' he asks as we are walking back.

'Yes, I think it is.'

'OK, I've got to dash to a meeting. Do please stay and enjoy lunch here, won't you?'

He speaks to the chef. 'Could you make Isabel a vegan lunch?'

I sat in the sun and they served me fresh juice. 'We make this from the fruit that is the wrong shape to prevent it from being thrown away.'

Then they brought beetroot that had been roasted on a very low heat and then put on a griddle like a steak and served on a bed of mashed potato, rainbow carrots that somehow tasted like caramel, tiny sliced fresh tomatoes that tasted as a tomato should, a pesto sauce that was made from the green leaves on the tops of carrots and peas cooked in their pods like mangetout.

And I think to myself while I watch the bees buzzing around the herbs, 'Yes, this is where I want to buy my food from and this is the way that I want to eat. And if it's more expensive, at least I'm caring for the planet. I'll save money on not shopping and not flying and I'll have joyful, tasty food instead. No more supermarkets for me. I'll pay organic farmers directly.'

If we all did this more, farmers would farm organically and this would have a major impact on slowing biodiversity loss. Guy's other recommendation: eat less meat and dairy – tick. And then there is 'ask difficult questions and be a pain in the arse.' I'm sure I can work on that one.

TRAVEL 1
Cars, Bikes and Legs

I have been doing some reading around the subject of air quality and cars. I can save you a lot of time. Your options, at present, are simple: 1. (The less-good option) Buy an electric car or 2. (The better option) Don't own a car. That's it. You're welcome.

To be honest, option 1 is very dodgy as you may want to consider where the electricity is coming from. And how much CO_2 it takes to make it. So, really it's option 2 or option 2.

There are an increasing number of people asking why, as a society, we accept the number of people killed and injured every year on the roads[3] as if it were some kind of natural phenomenon. This human just happened to be 'in the wrong place at the wrong time' as a huge metal box just happened to collide into them, ending their days a lifetime before they or their loved ones had intended. Sheesh! Talk about victim blaming. Pensioners, children on bikes, mothers with prams on the pavement – is it really 'just an accident' when these people's lives are ended prematurely by a car? Friends, do we really want these machines in any sane vision of the future?

I could tell you, in suffocating detail, about all the poisons that come out of fossil fuel powered automobiles, what it does to your lungs and how, for many, it's worse than smoking. I could give you statistics on the rise in asthma and other respiratory diseases, the long-term effects on health and the number of people a year that are dying and likely to die from these effects ... But I'll spare you the details. Just take my word for it. If you want to love the planet, private car ownership has to be relegated to the history books. That's it. We have to find life-enhancing alternative solutions because life is short.

There are little things that you can do to mitigate your car use. I'll list them for you: 1. Never use your car for short journeys. If you find yourself doing lots of short trips, sort yourself out so that you can do all the errands in one trip; if you have to do a school run, organize a carpool with other parents so that you don't all have to take a car out. 2. Make sure your car is in peak running condition with oil changed, tyres properly inflated, etc., and make sure you drive well because simple things like keeping to the speed limit and driving at a steady speed apparently make a small difference to fuel use. 3. Never, ever leave your engine running when you are 'idling' (this is the major factor in air pollution outside schools). Switch off your engine while you are waiting so that your car isn't spewing out fumes. But, friends, really all these are mere details. We have to be more radical – way more radical.

You may be thinking, 'This isn't possible for me. I need my car.' But I'm thinking of how our lives would be if, for some reason, we couldn't afford to run a car any more. And many people can't afford to run the cars they have. Apparently, in some of the wealthiest parts of the UK as well as some of the poorest, more and more cars are bought on hire purchase agreements and people are running up huge debts. Some car dealers will even admit that 50 per cent of their cars are bought on credit. It used to be just on houses that people put down a deposit and then paid it off forever, but in some places car debt is so huge some financiers fear it may have a major impact on the economy. It's a crazy world when we cripple ourselves with debt to own huge upmarket cars while others live very happily without them.

If you have disabilities, use an electric car and lobby to make sure the electricity being fed into the grid is renewable. For everyone else: let's consider how we can make the alternatives joyful. If you found you couldn't afford to run a car any more you would still go on living and working. You'd just get very creative finding ways to make your car-free life work for you.

'But how would I get my kids to school?' someone is shouting at me …

OK, so let's take that one first. Here is a favourite story to get you started. I have a friend with two children, a dog and a job. Every day she has to get her two children to school and then go to work. At the end of the day, she has to collect her two children and get them, and the dog, home again. She bought one of those bikes that has a little buggy on the back that the two children and the dog can ride in.

'It's absolutely changed my life.' (She's like an advert.) 'The journey from my house to their school is 3 miles and it takes 45 minutes by car with the terrible traffic and the parking, and it's grim. A friend bought us this bike and it's changed all our lives for the better in so many ways. The kids love riding in it and they didn't enjoy the car. We seem to make everyone smile as they see us pass. There are people we wave to every morning and we genuinely are a happy sight because we all enjoy it so much. And I feel good because I know I'm one parent who's not polluting the air around the school.'

'It has an electric engine which I can flick on if I need to keep up with the traffic or there is an uphill section where the bike becomes heavy, but mostly I don't use it as it's not hard to ride. I carry friends' kids in it too sometimes – they all want to ride in it. I get an hour's exercise every day before work and another hour in the evening – and it's good exercise because I really do enjoy it. I've lost a stone and been able to come off the anti-depressants I was on – and it's honestly all because of that bike.'

'You're not worried that it's unsafe on busy roads?' I ask, guessing that this would be the first objection from some people.

'There are cycle lanes. I cycle from Kennington to Clapham Common through busy London streets but the cycle lanes are so much improved. Most cities are really doing all they can to support cyclists. I've been doing this run for three years now. I don't feel in danger in the least and I've not even had any dodgy incidents.'

'Are bikes like this really expensive?'

'You can get them without the electric engine or second-hand for about £1,000, which is the same price as one of those posh Brompton bikes that fold-up. At the top of the range they are about £4,000, but I swear I've saved more than that by not using the car and giving up my gym membership. Really, my health has transformed and so have all of our lives. Because of a transport solution. Who'd have imagined that?'

The answer is, not enough people, judging by the mess of cars outside schools. Aware that this isn't a solution for everyone, I asked others that don't have cars how they get to work and to school. Here are some of the answers:

'We scoot!' Ah, scootering. Make sure to buy one for the grown-ups too. I still have vivid memories of running alongside my mother when we were trying to catch a bus (we were always late for some reason) and my legs not being long enough to keep up with her. There have been generations of weary parents begging, 'Oh come ON' to children who dragged their feet, as Shakespeare said, 'unwillingly to school.' But no more. Now hysterical parents have to shout 'Wait!' as their children scoot ahead of them and only have to be taught to stop at junctions. And everyone gets exercise. And they don't pollute the air. A scooter would have changed the quality of my entire younger self's life in the way the bike has changed my friend's children's lives and hers. I'd actually have enjoyed the journey to school.

Then there is walking to school. Or, as one parent told me, 'We skip. Sometimes people look at us as if we are mad but we don't care. We like skipping.' You may complain that in your case the school run is too far to skip, that your child would never skip, or that you'd look a little eccentric skipping, in your suit, to school because you are the headmaster. Personally I think it could catch on but, OK, if you insist that it's not your style and neither can you scoot in your suit, let's consider walking. Yes, walking. Some children and adults hate walking. I have a theory about this. I hazard a guess that the children who hate walking are the children who have been driven and expect to be driven. The adults who hate walking have become accustomed to their cars. They feel walking is somehow not good enough for them. What do you think?

There are many children – I propose – who have always walked to school and so it's just the way it is. They live at point A and the school is at point B. Some schools have very cleverly created 'walking buses'. A group of children, with a healthy adult or two, will start walking in a crocodile and walk all around the neighbourhood close to the school with children being added to the crocodile at each house they visit. Sounds good fun and healthy too. But even if there are no walking busses at your children's school, walking can be a seriously good choice.

One friend tells me, 'We raised two children with no car. We walked, we chatted, shared thoughts, made plans and looked at nature. Sometimes it was fun, sometimes not. But it was just the walk – come rain or shine. The walk was 20 minutes each way. When they were older and went to a school further away we gave them bus fare and said that they could take a bus or keep the money. They nearly always walked. Both of them still happily walk miles.'

That last sentence interests me. What forms our attitude to walking? Do you love walking or hate it? Personally I prefer to take a bus as I'm always in a rush to get to whatever I'm doing next. I always took a bus to school and a bus home. So I never looked upon walking as a valuable activity in itself. It has always been just a means to get to the next place. And not a very effective one because it's slow. But kids who are raised on walking, go on walking as it seems as they have learned – in their minds, bodies and souls – that walking has all kinds of benefits apart from just getting us from A to B. And now there are clever apps on your phone that encouragingly tell you, 'You walked 10,592 steps today.' Really, walking is the new driving.

If you are able-bodied, can take public transport and live in a city, there is simply no reason to own a car. A friend states simply, 'I went car-less three years ago. I use public transport most of the time and if I need a car I hire one. Life is so much easier. And cheaper.'

Can you imagine cities with no private car ownership? I don't hate cars. I can drive but I don't own a car so riding or driving is a treat for me. I take an Uber or a taxi sometimes, but mostly I ride a bike, travel by bus and by train or walk. I look wide-eyed at friends with cars, wondering how on earth they can afford to buy and maintain such an expensive and often unnecessary machine.

If you live in the countryside, miles from anywhere, the challenges are different of course. 'If I didn't have a car I wouldn't be able to see my grandchildren,' one friend writes. I didn't say this, but if this person didn't have a car and still wanted to see their grandchildren, they would have to move house to live within walking distance from their grandchildren, wouldn't they? If there was no private car ownership, what other solution would there be? And the positive side to this is that we would then have a community again. If we couldn't drive for an hour or three to visit the people we love, we'd choose, I suspect, to live near them. People would become less isolated and therefore less lonely. They wouldn't be forced to wander around anonymous out-of-town shopping centres because the local store wouldn't have closed because people would still use it. OK, I said above that I don't hate cars. I believe I've just changed my position: I do hate cars; they not only pollute our air but can be seen to be a contributory factor in human loneliness. And of course there are also the roads which carve up and destroy natural environments.

We are here to protect natural environments. So I return to where started: option 1. Own an electric car (which is OK if you have disabilities); option 2. Don't own a car. Option 2 is better.

TRAVEL 2
Flying Hypocrisy with Plastic Too

I hate airports too. I hate their insistence on turning passengers into consumers. I loathe the special pathway through a thousand perfume counters, the 'opportunity' to buy sunglasses, the expensive leather suitcases. I even dislike the food outlets with their plastic cups, plastic lids and plastic-covered food. I know I'm sounding a little ungrateful for the services airports provide, but when I go to an airport I don't want to shop for two hours, I just want to get on a plane. And of course I have mixed feelings about that too.

I try to be a radical environmentalist, but supporting the aviation industry by flying is one of the worst things I can do for the planet. I had thought that my simple decision to not take domestic flights but only the occasional international flight would satisfy me, but no. There is a certain mood that takes hold of me (of all of us?) when I know I'm doing something that doesn't accord with my values. And irritability is the result. I know that the plane will be spewing out gases: carbon monoxide, carbon dioxide, sulphur oxides, nitrogen oxides, black carbon, as well as various microscopic pieces of matter known as 'particulates' and even lead into the atmosphere. Then there is the noise pollution and the impact on the lives of the animals and humans that live under the flight paths – which is most of us that live in cities. I've done enough reading to know all this. But all that my reading has told me about what I could do to mitigate this is that my contribution to the pollution would be up to nine times worse if I travelled first class or business class – because it takes up more space. But travelling economy is the way most of us travel anyway. So airports put me in this grumpy state.

Campaigners are asking the government to introduce a carbon tax whereby everyone would get one free flight a year but after that they would be heavily taxed to discourage flying. This would mean that businesses would have to pay extra tax from constantly flying their staff around the world and party goers would have to pay more to fly across the world just for a long weekend. But people could still have one holiday with their families somewhere lovely and the world traveller could still take off with a back-pack but would be encouraged to stay away longer. This makes perfect sense. I hear the aviation industry is 'inching' towards this. But if we wait for laws to make us do the right thing we may have to wait a long time. If a carbon tax on flying was introduced I'd be OK as this is the only international trip I'll be taking this year. But I still feel incongruent and therefore in a bad mood.

An environmentalist going on a long flight halfway across the world to take a holiday in Sri Lanka just doesn't sit well. On the other hand, I'm going on holiday so I'm excited, but I still hate airports. I bring a flask of my favourite hot coffee, prepared before I leave the house. I sit sipping it pre-check in, like a grumpy troll. I buy nothing – on principle. And on top of all this, dealing with the habit of airport staff of never letting you know which gate your plane is at until the last possible moment. I sometimes wonder whether there is someone in an internal control room who delights in waiting as long as they can so that they can watch us poor passengers scampering along all those corridors with our cases of just the correct weight. Weight apparently is one tiny thing that you can do to lessen your impact. Travel as light as you can – a spiritual lesson from the aviation industry. It cuts the pollution in some small way because the plane needs less energy if we all pack as little as possible.

I don't like airports because they have become too much like shopping centres. Airports try really hard to sell me stuff I don't need. They make the whole experience one long battle against plastic. I glower at all the things that airports try to make me buy. A plastic blow-up pillow? Just for the flight? Can't I use my jumper for a pillow? 'Yes, I can!' I want to shout to some demon-god of capitalism who wants me to spend money all the time on stuff I don't need. Do I want to buy tiny amounts of cosmetics in plastic bottles?[4] No, I don't. I'm an adult and perfectly capable of transferring a small amount of shampoo into a little glass bottle and putting it in my suitcase. We don't need these to take abroad and pollute the places we are visiting. In Sri Lanka they have no recycling so, in desperate attempts to get rid of the stuff, they burn plastic waste by the roadside. If I'd taken these little plastic bottles and then left them in a bin in the small guesthouses we were going to stay in, this would have been what would have happened to them. That or landfill sites open to the air where they would have taken hundreds of years to break down. Is this what we want to take to the countries that we visit?

When I asked the writer and journalist George Monbiot what is the best action I could take to love the planet, he said, 'Bring down capitalism.' That seems quite a tall order. But if we all live simply and don't buy rubbish we don't need, that would be a step in the right direction. We have been told that consumerism leads to happiness. We're told it every day but we all know it's not true. And as we are discussing joy: yes, there is something perversely joyful about *not* shopping at airports. Perverse because it's so pleasing to walk blissfully through all those shops pushing purchases onto you and buying nothing. Nothing – that thing that all the great spiritual teachers want – is available at the airport. Please, join me.

Then there is the flight. Flying is a joyful and humorous exercise in plastic avoidance. First, you will need one of those special drinking bottles that have a filter inside so that you can drink water anywhere in the world safely. They are designed for hikers to be able to drink safely out of streams, but they also serve you well on a plane (drink your remaining water before passing through security and fill up again on the other side) because as soon as you sit down you'll be offered a glass of water in a single-use plastic cup.

My game is to say, 'No, thank you. Nothing in single-use plastic, thank you.' The stewardess today throws me a look that is a mixture of confusion and admiration. I indicate my water filter. Five minutes later they offer a blanket. To be precise they offer a folded blanket in a plastic bag. I smile sweetly. 'Nothing in plastic, thank you.' I indicate that I've had the forethought to bring a jacket for the flight. Then they bring around snacks. 'No, nothing in plastic, thank you.'

There is the question of headphones for watching movies. You may never have had reason to consider this, but I had a chat with the stewardess about it. When these systems were designed, most people didn't own their own headphones. But now many people do. And everyone has lost or forgotten to bring the aeroplane adaptor that is sold with some headphones. Why don't they sell these on planes instead of perfume, make-up and plastic aeroplanes? Aeroplane headphone adaptors for around £6? They might actually sell some.

Anyway, they don't, so everyone takes the cheap low-quality plastic headphones (in a plastic bag) that are offered on planes because we are led to believe that our own headphones won't work. It's not true. The only difference is that the headphones they provide have two pins and yours only have one. So, if you plug in your personal headphones, you listen to the sound of whatever film you want to watch through one ear only. If you take their headphones they will be collected up at the end of your flight and some very poorly paid worker sits, all day long, taking off the little bits of sponge that touched your ears, throwing them to the 'away' place and then replacing them with new bits of sponge. This worker then has to wipe every pair (I suppose you may have spat on them or dribbled in your sleep or something) and then put each one in a new plastic bag ready for the next flight. Please – don't take the cheap headphones. 'No headphones, nothing in plastic,' I repeat like a demented character in *Groundhog Day*.

Then of course there is the airport food. Today my friend and I are flying Qatar Airlines. Miraculously, the food is not served in single-use plastic but in hard plastic bowls that they obviously intend to re-use. Amazing. We can eat. But I go on with my mantra. 'Thank you so much but nothing in plastic, thank you.' I hand back the fruit juice, the cartons of 'fat spread' and other unidentified offerings. 'Would you like some wine?' The stewardess asks me, holding up a glass wine bottle. 'Yes, please,' I say, paying careful attention to what she is about to serve it in. She picks up a single-use plastic cup. 'No plastic, thank you.' I produce a carry cup from my bag. She fills it with wine. She laughs.

Five minutes later they come round with hand wipes. I know this is beginning to sound as though I'm exaggerating. But no. Hand wipes have plastic in them. They are essentially the same as the wet wipes (most of which still contain plastic) that are such a huge problem for our waterways, as people flush them down the toilet and they don't biodegrade. They are offering those ... wrapped in a plastic cover that you rip off and, er, throw 'away'.

'No, thank you.' I reply politely. 'Nothing in plastic. I believe that I can wash my hands in the toilet, is that correct? You have water and soap there?'

'Yes, certainly madam.'

I feel like saying, 'Then why the fuck are you giving me all this crap?' But of course I don't – she is too sweet and she doesn't make the silly policies.

'Then I'll wash my hands there, thank you.'

Then they come round with the hot drinks with plastic lids ... and on and on. Finally they go round asking for 'trash' and they are given bags of it. Everyone who has just accepted everything they have been offered can fill a bag with it.

Then something rather lovely happens. The stewardess walks up and says, 'You are a very special customer. We are all taking about you at the back.'

'I'm so sorry,' I say, 'Am I being difficult?'

'Not at all. We don't like the amount of rubbish that we have to throw away after every flight and you don't have any trash at all. We really appreciate that.'

Delighted, I was. 'Well, there you go. You're not an annoying pain in the neck after all,' says my travelling companion cheerfully. I waited for the stewardess to say, 'And so we'd like to present you with 10,000 free air miles,' and throw me into yet another ethical quandary. But she didn't.

So, friends, best of all – don't fly if you can possibly avoid it. But when and if you fly to go on holiday – if you are fortunate enough to be able to travel – please take a hiker's water bottle with a filter inside so that you don't have to buy single-use plastic water bottles wherever you go. Take your carry-cup and enjoy playing my 'avoid the plastic' game at airports and on planes. I know it's not much, and you can argue that compared to the amount of pollution planes create, this isn't worth considering. But there are a lot of passengers every day across the world on a lot of planes. And if we want to fly sometimes in order to keep our holidays, this is something that we can all do easily. Be lovely and kind to everyone, all the time, but be determined at the same time. OK? Don't take any plastic crap from anyone.

STUFF 1
The Life-Changing Magic of Not Shopping

Marie Kondo didn't invent decluttering. People have been removing stuff they don't use as a means of making life simpler for decades. But she certainly made it mainstream.

Many people love to hate her. She is a Japanese woman wrongly described as having OCD – which is disrespectful, both to her and to the people who genuinely have this condition. The media love to make fun of her by concentrating on the most extreme parts of what she teaches, such as rolling socks and not folding them so that they can 'rest' when not being used, thanking objects for the service that they have rendered you at the end of the day and just generally just being too clean and too controlling.

There was such a backlash against her strangely named book *The Life-Changing Magic of Tidying* that it gave birth to a book in response: *The Life-Changing Magic of Not Giving a F**k*. I hear that the TV series on Netflix (which I didn't watch) often did her further disservice by mostly showing the 'before' and 'after' situations and not giving time to the sometimes very difficult process of working out how we can bear to part with objects to which we have grown attached but which depress us. I still have dolls from my childhood that to me have entire personalities. They live in boxes. Why haven't I got rid of them? No – please don't write and tell me. Ha ha. I know!

So, to my mind Kondo's a genius. I believe that what she has done, accidentally perhaps, is take a complex spiritual process – that of detaching ourselves from objects – and made it simple. Not easy – but simple. She created the question, 'Does it spark joy?' as a very clear way to work out whether we want or need to give an object house space.

'And what does this have to do with loving the planet?' you may ask. 'Does it make any difference to climate change and pollution levels whether my house is in a mess and has too much stuff in it?' The answer is 'no' and 'yes'. No, because the piles of unworn clothes and unread books that lie around your house don't actually pollute the world. What they do is pollute you. Not literally (although there are more dust mites in a cluttered room so they do create indoor air pollution). But they pollute our minds because a messy, over-filled, over-crowded room creates anxiety. Everywhere you look there is a job that needs doing: a cupboard that demands fixing, a drawer that is overflowing and begging to be sorted, a tap that is dripping, books rammed onto bookshelves that you told yourself you'd read sometime or re-read sometime, dusty surfaces, too many clothes stuffed together into wardrobes that give you the feeling that you have nothing to wear, 15 mugs in the kitchen despite the fact that it's a house for 2. Sometimes couples in the same living space genuinely perceive that 'all the mess is his' or 'all the mess is hers'. Of course, to her all her possessions are valued items and all his mess is just 'mess'. He feels the same about her possessions. Families with children are often unable to walk across the floor as there are so many toys lying around. But of course the kids never play with them. Family coats fall out of cupboards despite the fact that each person only ever wears one or two. I've seen 18 pairs of wellington boots (I wish I were exaggerating) on display in a home of 4 people. People who live like this often feel overwhelmed and tired most of the time.

Years ago the artist and designer William Morris said, 'Have nothing in your home that you do not know to be useful or believe to be beautiful.' Gandhi said, with the greatest clarity, 'The less you have the more you are.' And then along came a Japanese woman who gave us a method to achieve this. I recommend her book – she encourages us to go through every item in our home and ask, 'Does it spark joy?' and if the answer is no, you thank it for the service it has rendered you and you take it to your local charity shop. That's it.

'But isn't this just creating more waste?' you ask.

In the short term maybe, as not everything that goes to your charity shop is re-sold. Some of it is just dumped – sent to landfill or incinerated at energy-from-waste centres. But what has changed when you return home from your charity shop visit to what feels like your new home, is you. A miracle starts to occur (the miracle she speaks of in the title of the book). You find that in having less you have become rich. Your wardrobe – formerly full of unworn clothes – now only has 20 per cent of what you had before. But you love all of it. Suddenly you have something to wear.

The bookshelves previously crammed with books that you didn't like or had no intention of re-reading are now filled with your favourite authors. Your kitchen is full of crockery you love, the surfaces are now clear for a vase of flowers and the vase is one that your grandmother gave you that previously had been in the back of the kitchen cupboard. You have given away so much and yet you are richer. You love everything in your home. The local charity shops love you, someone far away will benefit from the money you have raised for charity, and you have no desire to shop. Ever again.

This is why decluttering helps the planet. It changes our relationship to stuff. We start to want to live simply. People with too much stuff buy more stuff. People with very little stuff shop rarely and if they do, they know exactly what they want. There are no rash spends for those that know how many items of clothing they own. If you know that you have three cashmere jumpers, why would you buy another one? But there can be a semi-unconscious mindset (which I know well) that can't even remember what is in the wardrobe at home. Having wandered, half in a daze, into a shop, an item of clothing dazzles from a rail, looking colourful and cheap. Somehow it feels like an achievement if it ends up leaving the shop with you in a plastic bag. Expensive items are rarely bought mindlessly.

And it's the same with other items. If something in your home breaks and you find you can't fix it, replace it with one that is the least likely to break. The question: 'Will it last?' when buying anything is one we need to bring back if we're going to love the planet. With less and less, you end up not shopping. At all. And you treasure what you have because each and every item is something that 'sparks joy'. It's an item that you love.

And that is why Marie Kondo is a genius. You probably have a friend that has her book on their shelf. Borrow it. Or order it from your local library if you haven't already done this process.

Then join one of the many #buynothing groups on social media. You'll save a fortune, live a happier, simpler life and play your part in cutting pollution as there will be one less consumer out there. And leading a clutter-free life where you don't buy stuff leaves you time and money to enjoy experiences instead.

Every study that has ever been done on human happiness shows that it's experiences, not stuff, that makes us happy. Be more Zen – don't buy stuff. Try it for a year. Write to me when you're done. Please.

STUFF 2

Ethical Consumption: To Buy or Not to Buy

As environmentalists we have to question everything. Every tradition – every habit, everything we do. And, of course, everything we buy. I'm going to do a little experiment and consider washing-up liquid and laundry cleaner. When many people in the UK think of washing-up liquid, they visualize, out of sheer force of habit, a certain bright-green concentrate marketed as 'Fairy Liquid'. Some people even have a sentimental attachment to this brand as their parents may have used it – and even their grandparents. In 2021, 20 million people threw it into their shopping basket.

Fairy Liquid: the name itself makes it sound gentle, doesn't it? This memory of our childhood that smells the same as it always did. A picture of a baby in a nappy on the label. It has to be innocent, doesn't it? It's almost reassuring – like Daz, Bold or Ariel – products that many of us have grown up to trust just because they have always been there and, if they worked for our parents, why shouldn't they work for us?

Well, I have lots of reasons, you'll be unsurprised to learn. First, all four of the the above-mentioned products are made by the massive US company Procter & Gamble. Their CEO was paid 21 million dollars for the year 2022. We can reasonably assume that this wage has not been lowered since then. I don't know about you, but I consider that income a little excessive. I'm not sure I want to add to it. Also with even 5 minutes' research you can find a veritable collection of reasons you may not want to give them your money. Just put their company name into Wikipedia. You may like to read the section 'Child Labour and Forced Labour', or about the criticism of their testing on animals, or even the statement from P&G that they are absolutely not funding the Church of Satan.

But aren't all these companies as bad as each other? I can hear some exhausted reader asking me through the ether. No – some are very different. Some companies deserve your money and some do not. Search for the Ethical Consumer Magazine online – my trusted advisor (or large group of research advisors) on all things – and type in: 'Ways to save the planet when I'm going shopping'.

The brilliant folks at the Ethical Consumer (and I'm sure there are equivalent organizations whichever country you are in) have a chart. They tell you all the best makers of washing-up liquid and all the worst, giving them a score out of 100. At the time of publishing this edition, the company right at the top of their recommended best buys, with a score of 95 out of 100, is a company called 'Fill Refill', and second to bottom is – yes, you guessed it – Fairy Liquid. The only company below them is called INEOS. They score 4 out of 100.

So how do they work these scores out? you ask sensibly. And the answer is 'very carefully'. They contact all the companies and spend months doing research so that we, the consumers, can be completely aware of who we want to support with our hard-earned cash and who we don't. Because it can be confusing. As we know, there is a lot of greenwashing around and just because a company has put the word 'Eco' in the title of its product doesn't mean that it is guided first and foremost by ecological principles. In fact, in their red 'don't buy' list (I don't think they'll mind me telling you) is 'Ecover' and 'Ecozone', along with 'Persil' and 'Method'.

Now some of these products may be good in some areas – for example, 'Ecover' and 'Method' are both certified as cruelty free. But they still lose marks in that criterion as their parent company, SR Johnson, has no policy on testing on animals. Unless a product states clearly that it prohibits animal testing then you can't be sure. No one wants to imagine that some constituent part of a product has been put in the eyes of rabbits. I like to be careful. I'm sure you do too.

There is more. The researchers at Ethical Consumer ask the companies questions about not only how they treat animals but also how they treat humans. What country are the products made in? Under what conditions? Are the workers reasonably paid? Are the supply chains transparent? They look into the known tax conduct of the companies. They see whether they are transparent about that too. They ask all the questions they can about the company ethos – is it gender inclusive? Are they investing in oil companies and/or the arms industry? Are they shamelessly greenwashing by making vague claims about recyclability? Or are they, like the top company Bio D, creating so much solar power from panels on their own buildings that they are giving energy back to the grid?

They have a whole range of criteria for packaging. Some companies make a real effort to avoid plastic (in its many forms). Some don't give a … er … I mean, some don't. Friends – as I've said many times, we have to be radical as environmentalists. It's so satisfying to refill your glass washing-up-liquid bottle when it runs out rather than buying a new one. OK, maybe I need to get out more – but these are the kind of things that fill me with glee. Like sticking a finger up at the excesses of consumer capitalism. 'Just buying what I need' and not taking home yet another plastic bottle – it is a satisfying change.

There is also the question of which brands use palm oil. The horrible truth about cleaning products is that a lot of them contain palm oil. None of us want to contribute to the destruction of the rain forests. As individuals we can't do much. But we can avoid buying anything that has palm oil in.

The two remaining sets of criteria for getting a 'Recommended Buy' logo from the Ethical Consumer are questions about the impact on the climate. To do well, companies have to have calculated their total carbon emissions and worked out what they are doing to improve. This includes not only their own impact, but the impact that their products are having on your carbon footprint and mine. Have they created a soap powder that can be used at lower temperatures, for example? Can we use less of the product?

All this would be enough for me. But there is one final criterion that we may like to consider. How many harmful ingredients are there in the product? When it comes to how you wash your clothes, the chemicals most harmful to humans that are often used in detergents are phthalates, paragons, triclosan and even formaldehyde. Microplastic ingredients are also included under harmful ingredients, and some 'washing sheets' and often those with plastic covers that 'dissolve' also contain microplastics. The enzymes that are used in biological detergents have been linked in some studies to the rise in allergic diseases. Biological washing powder, which contains enzymes, and non-biological washing powder are not the same.

I can guess what some of you are thinking. You may be thinking: 'But Isabel, so many people don't care about any of this. They may be struggling financially and all they care about is "Which products are cheapest".' Or they may have very demanding jobs and be financially comfortable but time poor and just want to throw whatever is most convenient into the shopping basket. Or they may be so worn down by life that they believe nothing matters – it's the 'humans are going to destroy the planet anyway so the washing-up liquid that I buy won't make any difference' school of thought. I don't agree.

Maybe we are going to destroy the planet. But in the meantime I know a lot of environmentalists that are doing their best to save it. And 99 per cent of them are working in far more meaningful ways than I am. But my own small speciality is looking at all the myriad of changes we can make as individuals. I may not know much. But thanks to the Ethical Consumer I know the best washing-up liquid to buy. Their top three most recommended brands are 'Fill Refill' (score 95; a refill-based brand manufacturing in Northamptonshire who are both a B Corp and a Living Wage Certified company), SESI (a social enterprise company based in Oxfordshire, England) and Greenscents (score 89), the only company that guarantees that its products are organic.

So I've been talking here about washing-up liquid and laundry detergents, and we can put them together because they are mainly made by the same companies. The top makers of laundry detergent are the same as the top three producers of washing-up liquid. But with laundry detergent, the 'poor' and the 'do not buy' groups, according to the research from the Ethical Consumer, contain some very famous brand names: Comfort, Persil, Surf, Ecover, Method, Ecozone, Ariel, Bold, Daz, Lenor and – bottom again – INEOS.

My friends use some of these products. I'm going to try a little experiment in brand loyalty and human persuasion. Armed with the information above, I'm going to try talking to some friends, a couple, and see if I can get them to swap their brands. Does it count as 'preaching' if I provide information?

What do you think? If I can show that one product has a higher score for treatment of animals, treatment of workers, tax conduct, company ethos, packaging, use of palm oil, use of other harmful ingredients and impact on the planet – can I persuade anyone to change when they have such information? And how many will tell me to go away because what products they buy is their own business? Just how deep is the loyalty to Fairy Liquid?

So I try my little experiment and it half unfolds very badly (reasons for environmentalists to be discouraged) and half very well (reasons for environmentalists to be encouraged).

We'll call my friends Tim and Tina.

Having been read the information above, Tina says:

'Isabel, it's all very well for you in your vegan house where there is no greasy animal fat around. But I'm a vegetarian living with a meat eater – my first criteria for a washing-up liquid is how well it works. You haven't mentioned efficacy at all.'

It's true – I hadn't even considered this. In my experience all washing-up liquids work as well as each other. Apparently not.

She continues, 'We like the fact that Fairy Liquid is cheap, available at my corner store and I can slosh it around liberally.'

'But what about all the facts above?'

'Frankly, I wish you hadn't told me all that. I don't want to feel guilty every time I do my washing up.'

I sigh. This is evidently not my intention.

Tim says, 'Personally I'd have to test these things before I'd consider switching over. We've used other brands before and they don't seem to be as effective. What would make a difference for me is that I would put up with the product not working as well if I thought that the previous product was really harming my health and there was quantifiable data that could back that up – that would have the biggest sway for me. At the moment I'm blind to all the factors that you mentioned and the only criteria I'm considering is a product's effectiveness.'

'So efficacy is the main consideration for you?'

'Yes, and the possible detrimental effects to my health.'

The larger environmental impact of the product isn't on this person's list.

Tim goes on: 'To be honest I just don't have the bandwidth for considering all this. I mean – it's not just washing-up liquid, is it? It's everything. I can't look into all this before I buy anything. I have to focus on what is cheap and what works best for my convenience and that of my family.'

'But you don't have to research anything. That's what the Ethical Consumer does for you.'

Tina adds: 'I'm so busy, I barely have time to think. It's so convenient just to throw into your basket whatever's there. If there were two products and they were both as good and a similar price – of course I'd swap. I like the fact that Fairy is cheap and I feel I can use it liberally.'

'Neither of you mind that you are making a wealthy and questionable company richer?'

Tim replied: 'We all do that, though, all the time. Why consider cleaning products in isolation? It's not that I don't care about the other criteria you've raised – but I don't care enough to optimize my whole life around this. I have five things front and centre of my brain …'

'And which washing-up liquid to use isn't one of them?'

Tim's being honest.

'The things that directly affect me, I care more about. Why do it for one product and not for another?'

Tina picks up his point: 'I mean, you buy your mobile phone because it works best but it contains cobalt. Have you read about cobalt mining in the Congo? Or lithium mining in South America?'

'Can we please not change the subject to discussing phones?'

She says: 'But we aren't changing the subject – we are discussing efficacy. Would you change your mobile phone for a more ethical phone that works less well?'

'Have we established that other more ethical washing-up liquids work less well?'

'No, but would you swap your phone?'

'I'm sure that Ethical Consumer have done research on mobile phones and I will read it and study it and it will certainly be a major consideration next time I need to change my phone.'

'And would you take a phone that is more ethical but doesn't work as well?'

I start to feel annoyed. It's like, 'How can you call yourself vegan and still wear leather shoes?' Don't they call this 'letting the perfect be the enemy of the good'? Even if I'm a hypocrite for calling myself an environmentalist and owning a mobile phone – surely my hypocrisy isn't the point here?

I ask, 'Why are we discussing mobile phones? There is no comparison between the accursed mobile phones on which we are mostly all now dependent – and washing-up liquid – which is a much easier swap.'

Tina says; 'Unless you live with a meat eater, spend half your life washing up, and are on a budget.'

'OK. So what you both seem to be saying is that in your busy daily lives, efficacy comes as high as all the other ethical constraints when buying a product. Just as it is for me when I buy a phone.'

'Yes.'

'So if I could get you samples of the top three washing-up liquids on the list – those recommended by Ethical Consumer as 'Best Buys' and they work just as well – would you be willing to swap?'

'Providing the price wasn't absurdly higher, because, as I said, we use a lot.'

A mass-produced product like Fairy Liquid is always going to be less expensive than a product made by a smaller UK-based company.

Maybe I could try the 'supporting a small UK business' approach. 'Would you mind if it was a little more expensive but you knew your money was doing to a UK-based business and not to the profits of Proctor & Gamble?'

They looked at each other. 'Yes – we'd be willing to try them out.'

So this was hard work. This was the discouraging bit. If I want this couple to change their consumer habits, I'll have to do the work myself. Buy the products – take them to their homes and see what they say. Then when they run out, it will certainly be less convenient for them to wait for a 'Best Buy' product that I can buy in my local waste-free store than to throw the washing-up liquid in their shopping basket when they do their weekly shop.

This is a problem for small businesses everywhere. I have been using the Ethical Consumer's best buy liquid for some time and I much prefer it to any other brand that I've ever bought. I have long disliked Fairy, with it's sickly green colour – way too highly concentrated (I always feel sorry for the rivers, the seas and the fish) and the synthetic artificial fragrances that so poorly imitates natural perfumes. But I guess most of us just haven't tried alternatives. So how would we know if a small family-run business is making a product which is better in every way than our habitual purchase? And that we'd prefer it?

To take their point about efficacy on board, it's true that, last time I bought a mobile phone, I didn't look at Fairphone (the only phone brand with a 'Best Buy' rating from Ethical Consumer.) I didn't do the research. I just assumed that it wouldn't be as good as the phone I have and I don't know anyone personally who has one so I couldn't get a personal recommendation. But I'm sure you all know what they say about making assumptions? That they make an ass out of us. For all I know, the Fairphone 5 may be a better phone than the one I have. It's certainly a more ethical and sustainable phone.

'Another thing …' said Tina later. 'Surely ethical consumption applies to everything?'

'Yes it does. But it's not impossible. We could start by supporting the most ethical companies when we make a high-price purchase, couldn't we? An induction hob? A solar panel? An electric car? And then consider whether there is a more ethical alternative to the products that we buy regularly.'

'I guess so,' she said, 'and like you said, some swaps are easier than others.'

So this was all the difficult part. The part to make a joyful environmentalist need a nap.

Then there was the easy part.

'On the subject of washing powder,' said Tina, 'neither of us have any attachment at all to the washing powder we buy. On the contrary, now that we've looked at it, there are lots of ingredients that we'd rather not have in the house. So that's a straight swap for us. We'll take the "Best Buy" recommended by you and by Ethical Consumer Magazine.'

'That's great. It comes from Fill Refill, a company based in Northamptonshire and I can tell you where you can buy it locally.'

So that's the good news. Sometimes, where there is no attachment to a brand and you can just provide information, it can be very easy to change a consumer habit. And all these changes, over a lifetime – as I've said many times – I do believe they change the market slowly but surely.

Consumer choice is one of the most satisfying ways that we can make a difference and support those who are supporting the planet – every time we spend our money.

ENERGY 3

Burning Coal and Admissions of Stupidity

I have a confession to make. I have only the smoggiest idea about the sources of energy in my house. I have no understanding at all of how burning coal creates electricity.[5] I do know that the planet would be better off if we left the coal in the ground because burning it is a major factor in atmospheric pollution creating climate change – but don't ask me how. I have even less understanding of the 'natural' gas that runs my central heating system and enables me to cook. My knowledge of the energy I use every day, so crucial to the importance of the survival of our planet, is as clear as the London smog on a bad day.

I have to admit that I hadn't really made the clearest connection between insulating my own home, and therefore using less energy, with coal pollution and the air quality in our towns and cities. Like many people, I have been switching off lights and doing small things to save energy for years. But I hadn't completely joined the dots. My understanding has been similar to most people's understanding of how the human body works – we have some grasp of how the respiratory system influences the blood system, but few of us could write a clear essay explaining it.

But this morning I went to see a campaigner from Cleaner Air in London, Simon Birkett, and as we were discussing what individuals can do about air pollution, in what seemed to me to be a diversion from the conversation about air quality, he started to talk about home insulation.

Here am I writing a book about loving the planet and how to have an enjoyable and interesting time doing it, and it hadn't crossed my mind to look at the energy efficiency of my own house. Duh! What is wrong with me? Also, as I'm looking at what will be useful for all of us, this will help the planet and save us all money. Pondering why I didn't do this years ago, it strikes me that although I have chosen a green energy provider for both gas and electricity, I have never gone to any lengths to ensure that – as a household – we take as little energy from the grid as possible. Even though it would also save us money.

I have always assumed that making the changes that I'd need to make my home more energy efficient would be way outside my ability to pay for them. I know that a brand-new state-of-the-art electric boiler, for example, would heat my water and my house more efficiently than my ancient gas boiler, and not pump out emissions as my old boiler does, but I'd always assumed it would be completely beyond my financial ability to replace it. I've never actually found out what it would cost to buy and fit.

I have always just thought to myself (more not-joining-the-dots thinking – or lack of thinking), 'My system is inefficient, I'll just have to put up with higher bills.' And that the person suffering from this energy inefficiency was me –

and meanwhile I was going around taking trains instead of planes and refusing to use single-use plastic.

You'd have thought, would you not, that the very first step for anyone who wants to care about the environment is to make sure that they are living in the most energy-efficient house possible? Because – quite simply – the less energy we use, the less energy is required from the grid and the less pollution is created that destroys the quality of our air and our planet. The connection between someone throwing a cigarette butt onto the pavement and sea pollution is clear to me but somehow the connection between turning down the heating and air pollution hadn't been.

So I'm starting. It's possible to get the energy efficiency of your house measured. Normally people only do this if they want to sell or rent out a property. But you can also get yourself an energy-efficiency rating because you want to love the planet. As it happens, you'll also save money on your bills if you are able to do any of the work on your house that needs to be done. You'll even add to the value of your home if you want to rent or sell – but that may not be your primary motivation.

'Some houses are so energy efficient that you only need to put the heating on for a very short time for one or two weeks of the year.' Simon tells me.

I begin to feel ashamed of my old gas boiler struggling desperately to heat a property that is almost always cold and in which we sit in woolly jumpers looking out of single-glazed windows. We don't even have curtains. I got rid of the lined curtains that the property had when I moved in because they didn't please me. I never considered their skill at retaining heat in the room.

As an experiment, and to learn about the heating in my home, I've taken the simple step of switching off my central heating. It's February. I'm learning a lot. It's cold. I almost want to go and sit and work in my neighbour's house. I want to say, 'Excuse me, it's so cold in my house and I know you've got your heating on. May I come and work at your house?' And then I realized that there are seven people I can think of in my street that all work from home alone all day and all heat their properties. This in itself is insanity, isn't it? Why don't more people work in communal working spaces? If we had one of these, all the other houses could turn off their heating all day long.

Anyway, I'd better start with understanding my own energy use. An inspector is going to come to my house to test it and provide me with an energy performance certificate (EPC). This will explain everything that I need to do differently.

ENERGY 4

Gas and Hot Air

I got so confused today I was almost ready to believe every conspiracy theory you've ever heard.

A very young man came to my house to do the test for energy proficiency. I've been reading about climate change and I have, at least, understood that we have increasing amounts of renewable energy and that electricity is a cleaner form of energy than gas, even if it is currently more expensive. Gas burns fossil fuels which first, are not renewable and second, create emissions which pollute the air and by doing so accelerate climate change.

The government has targets to reach to cut emissions if we are to stand any chance of keeping global warming under the agreed target of 1.5°C. If the temperature rises more than this (which it is predicted to do), the consequences for us and future generations are apocalyptic. I won't go into them now, but let's just say that's it's not looking good for the planet. And my old gas central heating boiler has a flue which very much resembles an exhaust pipe so it's making the situation worse every time I switch it on. Not by much of course – I'm just one household. But so is every other household that this man visits daily. He works for a company called Vibrant, which is one of the many UK companies that offers an EPC (energy performance certificate) for your home. The testing system is devised by the Ministry of Housing, Communities and Local Government in consultation with the Building Research Establishment.

You can imagine my confusion then when the man told me that he was happy with my old non-condensing gas boiler and 'we will not be advising that you change to electricity'. I looked at him in amazement.

'But surely,' I asked, 'electricity is so much cleaner than gas? Are you not advising that people need to switch to electric?'

'Oh, no. Where there is an old electric system and a gas connection is available, we advise changing to gas.'

At that point I was ready to give up on my project and give up on the planet. Join those who think that our government is totally ruled by the oil industry; that our government doesn't believe in climate change at all … In short, I needed a very good red wine and was ready to drink a glass that had been flown in from New Zealand.

I rang every intelligent friend I have. Every environmentalist I could think of.

I left several garbled messages. 'This is like Pandora's box.' I felt as if not only had I opened the box but also that nothing flying out of it made any sense. Surely – *surely* – the government can't be encouraging endless households across the UK to sign up to gas just to keep the oil industry happy? Surely they must be genuine about their own targets? Surely the UK government believes in climate change?

Fortunately, an intelligent friend phoned me back. Rather hysterically I explained my confusion.

'Please don't tell me that this whole EPC scheme is just run by the oil industry to justify fracking or something.'

'Sit down and breathe, Isabel,' he advised. 'You are hopelessly confused.'

I sat down and calmed down.

He explained. 'The clue is in the name of the performance certificate. What they are measuring is the energy your house is using. Not the pollution levels.'

'Yes?'

'You are forgetting that, until such time as we have 100 per cent renewable sources for electricity, which we are a long way from having today, it still takes gas to create much of our electricity.'

'My service provider only adds renewable electricity to the grid. I only give my money to them.'

'Yes, but what you draw your energy from is still the central grid and a large amount of that electricity is created by gas. It takes more gas to create electricity to power your central heating than it does for you to use gas directly.'

'Er, wot?'

'So, what the government wants to do is just lessen the demand for all kinds of energy. So they are measuring your house for how efficient you are in making use of all forms of energy. They are less interested in changing you from one system to another than they are in cutting your overall consumption.'

'But what about the pollution?'

'It's not called a pollution efficiency test.'

'So what you're saying is that I may be doing less damage with my old non-condensing gas boiler than I would be if I had a new electric boiler?'

'At the moment – yes, that's possible. As more and more renewable electricity becomes available it will be less likely but at the moment, yes.'

'But surely, in this way they are creating more use of gas, and so undercutting the demand for renewables? So they can justify fracking for more gas because there is more demand. So the huge gas companies are still looking to their profits and this EPC report and the recommendations that come from it are helping them to do that?'

'There is a clear logic in what you're saying. I suppose the gas companies, which are private and run for profit, don't want to put themselves out of business. In the meantime, one solution would be for you to keep using gas but to replace your old boiler with a gas condensing boiler, which would use a fraction of the gas.'

'How much would that be?'

'Mine was two £2,000 but it pays you back in ten years.'

'I'll think about it. Well, at least I understand the energy performance test now. But it seems very short term thinking to me that they don't include any measure of emissions. You know what happens when windmills collapse into the sea?' (Bill Maher's Question)

'Tell me.'

'A splash.'

'It's true that we need to support renewable energy and this isn't very joined-up thinking. Anything else that was interesting about your report?'

'Yes. The part about roof insulation and wall insulation isn't relevant to me as I have another flat above me and I live in a terraced house so I have a neighbour on each side. But where my home would really benefit would be with double glazing. Some of my neighbours recently replaced all their windows and the bill was £10,000.'

'Oh dear. Yes.'

'I have secondary glazing on one side. This is the old way of making cheap double glazing – but the units aren't sealed and apparently it's not as good for retaining energy and excluding sound as modern double glazing. I thought I'd get good points for my secondary glazing, but no.'

'Do you have that all over your flat?'

'No. On the other side I have just single glazing so that would be where to start.'

'It would.'

'Thank you so much for explaining. It still seems crazy to me but at least I understand a little better.'

So if you want to increase the energy-efficiency rating of your house you need to do everything you can to use as little energy as possible.

If you rent your home then try asking your landlord if he or she will consider fitting double glazing. It will be in their best interest as it will add to the value of their property. It will help with your bills and it will help the planet. If you own your own place: ditto. The advantage (if you have any spare money at all that you want to invest in your home) is that fitting good-quality double glazing cuts down your bills and immediately increases your EPC rating, which is one of the things that will be considered when and if you either want to rent or sell. It's not cheap but it's a win-win and definitely a wiser choice for your bank balance and the planet than, say, buying a new car. Cars ... sigh.

'And is there anything else you can do?'

'I need to change all the remaining light bulbs that are not LED but I'm only going to do that when the current ones die.'

'And?'

'There is one more thing. They checked that my external door doesn't have draughts where the wind blows in. I have a very badly fitting cat flap. It seems I need a new draught-proof cat flap. He says I'm probably losing more energy through there than with all my old lightbulbs put together. And the neighbour's cats have an annoying habit of not closing the door properly behind them.'

'So they say you can keep your polluting gas central heating system but you need about ten thousand pounds of double-glazed windows and a new draught-proof cat flap?'

'That's about the result of today's research, yes.'

ENERGY 5
Smelly Emissions and Profit Margins

The EPC (energy performance certificate) report on my house arrived immediately. They had taken the various details about where I live and fed them into a computer. The computer had provided recommendations for the house to use less energy. As you might expect from a computer, it was not very logical. I had to spend an hour on the phone to the head of Vibrant, who had done the test. The man I spoke to was absolutely lovely, but I couldn't make sense of their report.

Their first recommendation was that I invest in ceiling insulation. As there is another flat above me, and they have both central heating and carpets, I hope you can understand me being a little bewildered by this. But even if I did this, the report tells me, it would only put my energy efficiency up by one point. When I enquired further, the recommendation for ceiling insulation actually relates only to one room, where there is not another floor above. This is logical but barely worth it because it's such a small space and it would only get one extra point. Also, the insulation would have to be 150mm (5.9 inches) thick to make any difference. The man who did the inspection didn't enquire as to whether there was already any insulation in the flat roof above. I think we can safely say that this test isn't the most efficient.

Their next recommendation is that I install internal or external wall insulation. This is also nonsensical because I live in a terraced maisonette so I have other properties on each side. But as for the other two walls, I've no idea what they mean by 'external wall insulation'. A form of cladding? I'm not sure any attempt of mine to add cladding on top of the Victorian brickwork would be much appreciated by my neighbours, even if it were possible or legal.[6] And internal insulation? Should I remove, for example, the cooker and all the kitchen cupboards, add internal insulation to the inside wall, and then replace the units? This, along with removing all the radiators, insulating the walls and returning the radiators, would cost up to £14,000, as they kindly let me know. If I did this then the EPC rating would be raised by up to seven points.

Changing all the lights in the house to LED lights (at the moment apparently 56 per cent of the lights are LED) would also put the rating of the property up by one point. Replacing all the single-glazed windows with energy-saving double-glazed windows costs between £3,000 and £7,000 – but I'd be moved up only two points.

This is perplexing information. Their key requirements are not logical. I may look into 'exterior wall insulation' just for fun. It may be interesting to see whether it would even be possible to get local government permission to do what the government recommends. Surely people can't just go sticking stuff on the front of old buildings, even if it's not a conservation area? And if it's not possible then why is the government recommending this?

But it gets stranger. Unbelievably, I've got excellent marks for that ancient and highly polluting gas boiler. They even noted details of the model number. It turns out that if I ripped out the entire system and replaced it with a new, modern, state-of-the-art electric central heating system (a very expensive option) which would produce zero emissions – my EPC score would go down.

'Down?' Heaven help the planet.

'Considerably,' the man explained to me.

This is inexcusable. They are more concerned with saving energy than they are in preventing emissions. So with my old system that pollutes the skies daily I have a higher grading than someone producing zero pollution. I'm not making this up. The reason – as I explained – is that they are only concerning themselves with the amount of energy you use.

They write at the bottom of the report:

'The average household causes about 6 tonnes of carbon dioxide every year. Based on this assessment, your home currently produces approximately 3.6 tonnes of carbon dioxide every year. If you were to install these recommendations you could reduce this amount by 1.7 tonnes per year. The higher the score of your property, the less impact it has on the environment.'

'But you don't need to get anything done at all,' the 'energy' man explained to me. 'Your house is at the top of the D band with a score of 67. That's way above the national average.'

'It seems mainly to have that score because I'm running an ancient boiler that doesn't work very well,' I replied cynically. 'Your man should have taken one look at my system and told me that if I wanted an energy performance certificate of any kind it would have to go.'

'It's not our place to assess whether a boiler works well or not. We simply take a note of what kind of boiler it is and feed it into our algorithm.'

'Very logical.'

'It's fair to say that this system is deeply flawed,' he said honestly.

I was struggling to take something positive and practical from this experience. 'I notice that although you don't want me to go electric, your report would like the idea of me changing to a combination boiler or a condensing gas boiler.'

'That uses less gas.'

'So you would encourage me to remove an old electric system and change to gas?'

'Where a gas connection point to the mains gas exists – yes.'

'And, from an environmental perspective you can understand me being surprised by that recommendation?'

'It's fair to say that the government is currently pushing gas.'

I ask this question again as I can't quite believe it.

'So you'd rather that I replaced my system with a gas condensing boiler than a new electric boiler, even though the long-term policy is more renewable electric energy?'

'That's correct because we are interested in cutting your energy use and a condensing boiler is currently the most efficient way to do that.'

'I see. I have one more question. Is the government insane? Are you insane? Or am I insane? Are you actually being paid by BP or Cuadrilla?'

(No, I didn't really ask this. The man I was speaking to had one of those wonderful Thomas-the-Tank Engine Welsh accents. It wasn't his fault. I was obviously going to have to figure out how to make my house more energy efficient on my own. But I'd like to get some more points on the absurd government algorithm too. Almost just for fun. I'm determined to find some benefit from their test.)

'If I were to insulate my kitchen ceiling and replace all my lightbulbs – according to your chart I'd be two points higher and that would put me into the C band even if I didn't replace my single-glazed windows. But presumably I'd have to be retested?'

'That's correct.'

'Is it possible that a different tester could come to a different conclusion?'

'It is.'

So, my conclusion and recommendation so far, is this:

Since we are exploring what we can do to help the planet that is also wonderful, long term and positive, I would reach a very different conclusion from this test. I'd say, do upgrade your windows from single to new super-good-quality double glazing if you can. It will help you keep warm, save you money on heating bills, increase your EPC rating and the value of your house, keep out noise pollution and look fabulous.

Unless you are building your house from scratch, do not even think about external wall insulation. Internal wall insulation seems equally impractical unless you are ripping out kitchens or bathrooms, in which case you could add internal wall insulation before refitting – if you have the budget. Ceiling insulation seems more possible and cheaper, if you don't have anyone that lives above you – and I'd explore the cost of this along with solar panels too. This will save you energy, cut your bills, save the planet and even meet the criteria of the current government's EPC.

Change your light bulbs, as I mentioned, but only when they blow. But do not – under any circumstances – replace your electric central heating system with gas. Because, friends, if you are cooking and/or being heated by electricity then you are ahead of the rest of the population, not behind it. So, you just keep your electric system. Make sure that you pay money only to a company that puts renewable energy into the grid.

I met a fantastic campaigner last night who lobbies for no private car ownership in London. She's quite the environmentalist. She has downsized from a four-bedroom home to a one-bedroom studio flat to be sure of using less energy. And she said it took her two years to decide how to heat her space …

My advice would be different from the government's. I'd say, go electric if you can afford it and just wait for the EPC rating process to catch up with you. They will be forced to include emissions soon. So you'll be ahead, warm, happy and pollution free at your end. And as they add more and more renewable energy to the grid, you may end up being pollution free at the supply end too. And by adding to the demand for clean renewable energy, you'll be encouraging that to happen faster, too.

If you want to have a reasonable EPC rating, you're just going to have to do something completely strange like insulate your ceilings. And I don't think adding a green wall of plants to the outside of your property counts as adding external wall insulation. Not yet anyway. But you could try. Or fit external wall insulation and put a green wall on top of that. And add some wonderfully lush, thick, gorgeous, colourful curtains to keep the warmth in during winter. I'd spend as much money on those as you possibly can.

WINE 1
Cosmic Grapes

Buying wine. Someone has to. It strikes me that I've been doing this wrong for many years. I've been looking at rows of bottles thinking, 'Wine from Chile? That sounds wonderful. Wine from Australia? Why not? There's lots of sunshine in Australia. Wine from New Zealand? South Africa? Why not?'

I'm sure one of my discerning criterion has been, 'I love the design on that label.' Or even, 'This must be good as it has a sticker on the label that says it has a gold medal.' I know, I know – this isn't how you're supposed to buy a wine. I hate this. This is like buying books just because they are on the front table of the bookshop in the buy-one-get-one-half-price selection.

I can save you time. If we are going to buy wine and save the planet we need to buy organic wine. And preferably 'biodynamic' wine. Biodynamic is even higher on the hippy scale than organic. Not only do these winemakers avoid all chemicals, pesticides and artificial fertilizers – as organic farmers do – they dance naked among the grapes in the moonlight. Well, almost.

Biodynamic farming predates organic farming. It was created in 1924 by Rudolf Steiner, who is now known mainly for creating schools that children actually enjoy attending. He was a mystic as well a philosopher and an ecologist. Not only did he advocate farming in a way that 'restores, maintains and enhances ecological harmony', he also suggested that humans have a responsibility to their environment that isn't based exclusively on profit. Can you imagine such a concept? His work had a spiritual focus as he believed that care of the soil, as well as animal and human welfare, were all the responsibility of the farmer. To Steiner, a farm was a living organism with its own individuality. He believed in Mother Earth, was a vegetarian and even liked humans too. Biodynamic also includes a commitment to community, saying that the produce should be sold locally wherever possible.

While Steiner taught that the earth is a living entity, he also said that certain actions needed to be taken to release her 'cosmic forces'. Some of them read more like spells than scientific farming methods. The ground was to be sprinkled with various herbs and planting was to be done to coincide with phases of the moon and aspects of the zodiac. So, you can either believe that it's logical to plant in conjunction with the moon – because some phases have more light than others – or you can dismiss him and many of his teachings as pseudoscience. In other words, as having no validity whatsoever, as many have done.

However, in 2018 the Demeter Association in Germany came up with a very strict list of criteria that needs to be met for a farm or a vineyard to qualify as biodynamic, which even includes post-harvesting handling and processing procedures. And here's the bit you want to know:

In a blind testing of ten biodynamic and conventionally grown wines, judged by a master of wine and head sommeliers for *Fortune* – the investor's magazine – nine out of the ten biodynamic wines were judged to be superior. Isn't that satisfying? Furthermore, biodynamic practices have been found to be better for the soil as they are required to leave a minimum of 10 per cent of the land for nature. And the farming practices are more energy efficient too.

So, what I'm leading to is this: if you want to love the planet more you need to buy more biodynamic wine made in France or Germany. Failing that, we can buy organic from Italy, Spain or elsewhere in Europe – but we certainly don't need to have our wine shipped halfway across the world. And is organic English wine any good, you ask? Well, I think we need to find out. We are environmentalists after all. We are learning to love the planet. It's the very least we can do.

ANIMALS 1
Myths and Palaeontology

Has it ever struck you as interesting the amount of dinosaur products that are marketed to boys and unicorn products to girls?

You can't get away from dinosaurs in the UK. I recently visited the wonderful Horniman Museum only to discover that it had been taken over by 'Dinosaur Evolution', an exhibition about extinct creatures. There is currently a touring event too. We are being offered a 'World of Dinosaurs' which features: 'thirty, moving, snapping, roaring dinosaurs'. The website boasts: 'From cunning Velociraptors, to vast Brachiosaurus, from terrifying Spinosaurus to unpronounceable Pachycephalosaurus, plus of course, a mighty T Rex, they are all here!'

The Natural History Museum offers the bones of a dinosaur. (Actually they are replicas but they are at least replicas of original fossils.) There is 'Dinosaur World: Live!' – a piece of theatre returning for 'Another roarsome summer season'. In Norfolk they have 'Roarr' the 'UK's largest dinosaur themed park'; in Torquay, 'Dinosaur World'; on the Isle of Wight, 'Dinosaur Isle Museum'; in Dumfries in

Scotland, 'Dino Park'; in Swansea, 'Dinosaur Park'; and in Belfast the Ulster Museum features, yes, dinosaurs.

In Crystal Palace Park in London there is a collection of clay and cement dinosaurs that have been there since 1854 but only three of those resemble true dinosaurs. I was there recently and I heard a young child say, 'Daddy, I saw a crocodile's tail!' No, sadly, what she had actually seen was a piece of cement modelled to look like a creature that wasn't there and never had been. To compensate, the park has a lot of pigeons. And I saw some coots.

'But my son and daughter love learning about dinosaurs,' I can hear some reader complaining. OK, but hear me out.

Unicorns – marketed exclusively to little girls, usually on products that are also pink as it's a colour we are told that little girls love – are a 'symbol of purity, grace and innocence'. Their horns were said to be able to heal people and purify poisoned water. 'But is it true?' – well, you guessed it, no. There is no evidence that unicorns ever existed in any country anywhere in the world. Dinosaurs aren't around either.

So is it a coincidence that so many little boys read about dinosaurs and so many little girls dream about unicorns?

It's almost as if somehow, someone somewhere doesn't want our children to become interested in animals that are actually alive. I'm not a conspiracy theorist and I'm not saying that anyone has sat down and done this deliberately. But how many of our children are going to want to grow up to be naturalists and environmentalists if they aren't interested in real animals?

I know children that can list ten different types of dinosaur but don't know the difference between an African elephant and an Indian elephant. I know children that don't know that unicorns don't exist, but neither do they know that on

Dartmoor up to 2,000 Dartmoor ponies still live wild. They go to bed hugging a toy unicorn but have never sat on the back of a horse or a pony.

So parents, grandparents, aunties, uncles, brothers and sisters – please introduce our tiny humans to the world of living animals. Adults must also be boring themselves silly talking about stegosaurus. It is just me? I really don't understand why palaeontology would be included on the list of subjects I'd want to introduce to a junior-school-age child. But the parents may be interested in the fact that across Europe brown bears are being successfully reintroduced?

If little Jonny or Suzie learn all about bears, then, if they are very lucky, perhaps their parents may find a way to take them on a trip to see the object of their interest living free in the wild. And it doesn't even have to be bears. Beavers are fantastic little mammals to learn about that have gorgeous tails, and build little dome-shaped dens to live in called 'lodges'. Beavers gnaw down trees, build dams, re-route rivers, create whole new ecosystems and prevent the flooding of human homes. Even in the UK, there are more and more places where you can take Jonny and/or Suzie and teach them that, if they are patient, they might see one swimming along with a stick.

There are naturalists now, in the UK, who hope to see the reintroduction of the beautiful grey wolf to our shores. People are afraid of wolves because they know so little about them. But there is a role available, in the future, for the first man or woman to successfully reintroduce a pair of wolves to Britain, just as has been done across Europe. Maybe that person is alive now – a child. So if you are their parent – please don't give him or her a book about dinosaurs. Surely wolves are more exciting?

If they want monsters – tell them about the Gila monster who lives in Mexico and the southern USA. If they want magic, tell them about chameleons who are alive now in Madagascar and change colour. Tell them about beautiful peacock spiders whose eight eyes are so big that they take up the whole of their heads. If children become curious about insects they won't grow up with a fear of them. If you show children how extraordinary and intelligent octopuses are, when they are adults maybe they won't want to eat them. If you tell them stories of the orange oranutans, maybe, as adults, they will want to visit Borneo and play a part in protecting the rain forests. Maybe they won't be ecologists but at least they may want to avoid buying palm oil.

If you would like children to fully enjoy nature and care about animals, introduce them to meerkats (still living in South Africa and looking out for everything) or the lazy brown-throated sloths of Central and South America, forever taking naps. Or huge sea turtles hauling themselves onto the sand to lay their eggs on the beaches of the Atlantic and the Indian oceans. If you want to start nearer home and are desperate to move beyond cats and dogs, maybe you know someone who keeps chickens? Or has hedgehogs in their gardens snuffling around at dusk?

I'm basically begging you – all those that are looking after the next generation. Put a picture of a bottlenose dolphin on Junior's bedroom wall, a moose, a porcupine, a gorilla or a penguin. If you want a creature that looks prehistoric choose a pangolin. If you want something with a 'horn' choose a swordfish.

If we want to take care of the joyful environmentalists of tomorrow please choose any book, any lunch box, T-shirt, fluffy toy or 'experience' that is not about dinosaurs or unicorns. Thank you.

CLIMATE CHANGE 1
Flat Earthers and Climate Change Deniers

Global heating? Climate change? Climate breakdown? Confused? Who wouldn't be? Our confusion has been created deliberately. The climate change deniers are mostly politically motivated. According to Naomi Klein's book *This Changes Everything*, they are a very small group funded, in one way or another, by the petrochemical industry and there are very few actual scientists amongst them. In the case of the 3 per cent of scientists that are going against the other 97 per cent, their work has been found to lack scientific rigour of any kind. What you can be sure of is that climate change is beyond reasonable questioning. Or beyond questioning by reasonable people.

The job of the active-denier lobby is to make it appear that the question, 'Is climate change a reality or not?' is equally weighted. It's like asking, 'Did the Apollo spaceships really go to the moon or was it all rigged up in a TV studio?' Or 'Is the earth round or flat?' You can find people who believe the earth is flat. They are out there. And if they were being paid by the petrochemical industry then their views would be all over social media. They would be lobbying lawmakers to prevent people who want to prove that the planet is round from being listened to.

My friend Dr Jenny is a scientist who understands climate change. It's always a good idea to talk to someone who has spent a lifetime studying a subject rather than someone who has an opinion based on a few conversations at the pub. Especially if the uninformed opinion is convenient to them (because they work for an oil company, for example). Anyway – as I write this Jenny is also seven months pregnant. I phone her up to congratulate her. And then I ask her some questions about climate change. As you do when you're rejoicing with an old friend who is about to have a child. I hope I've asked her all the questions that you would have asked if you'd been on the phone.

'How are you? How come you haven't announced your news?'

'I'm going to the Arctic next week on a climate change project. I thought I'd tell extended friends on social media with a photo of my bump taken in the Arctic. It would be wrong not to.'

'Absolutely. But I'm encouraged by this news. You think there is a future of some kind then?'

'Climate change is happening and I'm having a baby nonetheless.'

'This is very reassuring. This is doubly good news. And you're taking your bump to the Arctic? Is there any Arctic left?'

'The permafrost is melting at higher latitudes. We are going to see what's happening in Spitsbergen. The land isn't as stable when there's no ice around so they have a lot of problems with coastal erosion, landslides and other very serious and urgent problems. We're going to see how the people up there are coping and some local authority representatives are coming to learn how we can educate people back home.'

'Please remind me, what are you exactly?'

'Apart from being a very pregnant woman who often feels very tired? Hmmm. I'm mostly a satellite data scientist. Like it or not, I can confirm that global warming is happening. I have studied the data, both from satellites and from climate models and the data back up the model results.'

'Why do they call it "warming"? I mean, "warm" is such a comforting word. It makes it sound as though we are just going to get more warm summer days and lower heating bills.'

'The term "greenhouse effect" was first used over 100 years ago when scientists first described the warming of the planet which could happen as a result of burning fossil fuels.[7] But – just to be clear – it's not a good thing.'

'Can you tell me, simply, in words a non-scientist like me can understand, exactly what global warming is?'

'It's the rise in the temperature of the earth's atmosphere. Global warming can seem simple but climate change is much more complicated because our atmosphere, and the whole earth system, is very complicated and highly interdependent.'

'I have one friend who believes that none of it is true. She tells me that carbon dioxide is part of the natural atmosphere. How do I reply to a climate change denier?'

'Well, she's right that CO_2 is part of the atmosphere – we breathe it out. The trees and plants absorb it and create oxygen and that's part of a cycle that has existed since plants first evolved on the planet. But what we have seen since they started taking measurements in 1958 is that the trends in CO_2 now are not natural trends.'

'That's not just in places where there are too many cars? Surely there are places where there is the right balance of oxygen to breathe? In the middle of an ocean or something?'

'If you look at the measurements from Mauna Loa in Hawaii, which is the middle of the Pacific, so not directly affected by industrialization, you see the seasonal cycle, which is the natural variability, and on top of that there is an upward trend in CO_2. There are other things that can cause variability, but there have been so many studies done linking the increase in CO_2 and other greenhouse gases to human activity that it's not questionable any more.'[8]

'Some people just don't believe the studies though, do they?'

'There are more and more studies not just confirming what has been suggested, but also showing that many impacts which were less well defined in the past are in fact contributing significantly to climate change. And there are more and more studies about the research, showing that not only do most studies support climate change as an active and urgent global issue, but also that those studies that showed the contrary were based on erroneous facts or assumptions. And that's not even taking into account the work of the IPCC, for example.'

'The Independent Police Complaints Commission acts on climate change?'

'No, no. The Intergovernmental Panel on Climate Change. It was set up in 1988 by the UN and the World Meteorological Organization to provide a scientific consensus for policy makers to act on. Each chapter of their reports employs a panel of experts who reviews the studies of hundreds of authors for each climate-related topic. Each chapter covers a different aspect of the climate: land, ocean and air, for example, and based on their reviews the panel comes up with an overview of the scientific knowledge and the level of uncertainty for each aspect raised in the chapter. This way, with each new report published every five to ten years, scientists can focus on the areas where we have least knowledge.'

'How many scientists are involved in this?'

'For the report that came out in 2022, there were 722 authors. If you were to look this up,[9] you can find the panels that have reviewed all the previous reports. It's an astounding achievement because, well … trying to get scientists to agree on anything is worse than herding cats. And they do this to try to influence policy makers – to let them know that something serious is happening.'

'Why is everyone so confused then? If the planet is getting hotter, why has someone decided that we can't call it global warming any more?'

'Because it's too simplified. The planet is warming but our atmosphere is complex. For example, in Northern Europe we can expect to have more frequent and more intense storms as a result of climate change. As the atmosphere gets warmer it also gets wetter. We have already seen the growing season expand, so spring comes earlier and autumn comes later. But insects and plants have evolved over hundreds of thousands of years and their mutual timing has obviously evolved over this period as well. So, if you change the timing of the plants, the insects have to adapt as best they can, which has an impact on our entire ecosystem. Just saying, "our temperature is going to rise by one degree" doesn't cover the variety and severity of impacts that we're going to experience.'

'And yet everyone goes on driving cars, watching TV and getting excited about who wins football games?'

'Denial isn't just da river in Egypt.'

'I like that.'

'You're welcome.'

'So what do we need to know?'

'Emissions caused by humans have created global warming which has led, for example, to there being a drastic decrease in sea ice in the Arctic over the past 30 years. If the Arctic ice goes on melting at the speed at which it is melting now, we are going to see drastic changes in the European and northern latitudes' climate.'[10]

'What's your job title again?

'I'm an earth observation scientist.'

'And one of the things that you observe is climate.'

'That's right.'

'Last week we had a warm spell and one of my neighbours said to me, "If this is climate change, let's have more of it." There even still seems to be confusion about what is weather and what is climate.'

'Climate and weather are closely related, but are not the same thing. Climate is the mean of the weather conditions; weather is the short-term noise around that mean. So, as climate changes, weather changes and vice versa. This is why we don't talk about weather trends. They are different perspectives on the same phenomena.

'Weather is the short term and climate is the long term?'

'If you insist on simplifying it even more – yes.'

'So, is there anything particular that you suggest we can do?'

'Yes. Sort out your energy use. Don't be part of the problem. And become a little bit activist – the sooner the better. Business as usual is no longer feasible without serious consequences: let your politicians know that you know this and that you care.'

'A little bit activist.' I like that phrase. It sounds easy but powerful all in one line.

INSECTS 1
Bug Houses and English Bluebells

Insectageddon is a long word. It's an accurate one, though, as our insect populations are rapidly disappearing. I remember bug splat on car windows after a drive. Now we don't have bug splat because we don't have bugs. Ask any elderly car owner. We have so few left that we don't need scientists to tell us that insect populations are in a mess. According to some sources the numbers are going down so fast that, at current rates of decline, in 100 years we'll have none left.

And why would this matter? Some readers are probably thinking, 'I don't like insects anyway. They are like the Romans (think Monty Python): What have they ever done for us?' They're 'gross' (according to many city dwellers unaccustomed to anything with more than four legs); 'They eat my plants and they bite or sting me'; 'They are small, horrible and squirmy.'

The Romans gave us roads, sanitation, irrigation, medicine, education, peace and wine. Insects give us life. They nourish the earth and help to decompose wood and anything dead. It strikes me that some city dwellers, who, like myself, have spent far too much time on social media and not enough in the countryside, may actually not know this stuff. Insects are also food for other insects and mammals. Even in back gardens they are food for hedgehogs, mice, moles, lizards, frogs, birds and fish. And what about my most-disliked insect, the mosquito? Well, first of all only the females bite and then only when they need to lay eggs. And even the mozzies are food for bats. Bats have to eat too. Of course, a wide range of insects (not just bees and wasps) pollinate our flowers and our crops. Without insects, birds and animals would starve and eventually the entire ecosystem would collapse and we'd all starve.

Our planet would benefit from having far fewer people and far more insects. So I look up 'How can we preserve our insects?' An online search suggests 'in amber', 'on a pin board' or 'in hand sanitizer'. The entire discipline of entomology seems to be mainly based on trapping and killing insects in order to study them. I once visited an entomology department and was informed that there are no ethics whatsoever in the study of insects since they are not considered 'conscious'. Really – as I often say – I'm not making this up.

'Is there any way that you can study insects without killing them?'

'No,' the head of the course insisted. 'Entomology students have to be able to identify them and if they don't trap them to learn about them, they won't be able to identify them.' The room was full of insects in jars, insects mounted on pins and dead insects being looked at under microscopes. It just didn't seem very … well … kind.

So, I'd like us to consider how – in our own small way – we can love and support our insects better and keep them alive. Make our gardens, parks and farms buzz again, which will help preserve our bird populations at the same time. There is little we can do immediately about farming and agricultural policy. The best things we can do to influence this is only buy organic fruit and veg. Cheaper fruit and veg is usually not organic and so is sprayed with all manner of chemicals to kill off everything that moves, including – eventually – you. But our point of power here is our gardens.

In the UK it's estimated that gardens cover 1 million acres. That is an area larger than any of our national parks. If you don't have a garden you may have a back yard or a window box or know an elderly human with a garden who would welcome a little help and company. If you have access to any earth, anywhere, you can be part of the solution. And this really is a joyful win-win as it gets us away from our screens, and anything to do with plants has been shown to be one of the best activities that any of us can do to support our own mental health.

It turns out that we have been gardening all wrong. For years. We have been gardening for aesthetics. Duh. We are so human-centric that the only thing most people have cared about when considering their gardens is what they look like. And in some cases the showier the better. So we have imported plants from across the world because they are visually appealing – but they are bugger-all use to our native insects. Some of them have no nectar and no pollen and some of the double-headed varieties of these plants are even impossible for the pollinators to access. Can you believe how dim we are?

There are two kinds of garden that are really not helpful. One is a large square of nothing but grass with maybe a solitary tree or two in the corner. If there are no flowers or other plants, how are insects supposed to find somewhere to live, have sex and reproduce? A patch of bare grass isn't a garden – it's a barren desert. The only thing worse than that is artificial grass that's made of plastic. Don't even get me started on that abomination. They have some at the Chelsea Arts Club where I was taken once as a guest. I stared at it in dumbfounded disbelief. They are supposed to be a bunch of intelligent, artistic people that understand aesthetics, culture and how the world goes around. Plastic grass?

Don't tidy up your garden. If something dies – anything – leave it alone. Insects are there to help decompose whatever they find and that process nourishes both them and the earth. If you are pruning your trees, leave the tree logs on the ground. We have so little rotting tree debris in our garden, I'm seriously considering scavenging some next time I'm in the woods. This would probably be illegal and might well kill whatever was living on the rotting tree stump, though. Better to have decomposing wood in a garden. If you have a dying tree, let it die. Don't tidy it up. If you have fruit trees, leave the fallen apples, pears or plums on the ground and in late summer butterflies such as red admiral and painted lady will feed on the juice. Basically, anything rotting is good news. Honestly. Learn a new word: detritivore. Isn't that just the best?

This next bit is utterly joyful: insects, just like humans, like nature to be full of variety, colour and perfume. In a study that the Royal Horticultural Society did at Wisley, over four years, they found that insect populations were higher when areas were planted with mainly native British plants with some semi-natives and non-natives to extend the flowering season. The study was called 'Plants for Insects'. In case, like me, you have no idea what is native and what is non-native, here is their list of native British plants. Ideally we need to have up to 70 per cent of plants that are native to the UK in our gardens:

1. Sea thrift
2. Common box
3. Common broom
4. Tufted hair grass
5. Maiden pink
6. Male fern
7. Hemp agrimony
8. Bloody cranesbill
9. Common rock rose
10. English bluebell (not the Spanish ones apparently)
11. Field scabious
12. Ox-eye daisy
13. Common honeysuckle (Graham Thomas)
14. Purple loosestrife
15. Purple moor-grass
16. Primrose
17. Sweet briar
18. Small scabious
19. Betony
20. Common valerian
21. Spikes speedwell
22. Guelder rose

Isn't that a splendid list? It doesn't mean that there is anything wrong with most of the other plants in your garden. Of course not – you don't have to go pulling them all up. It's just that if you are going to buy plants or collect seeds to grow your own plants then these would be good to include. Today I went out and bought some English bluebells. Absurdly satisfying.

Here are a few other things that we can do that are well worth enjoying. Next time you spend a day with some kids and don't know what to do with them, look up 'How to make an insect (or bug) hotel' on YouTube. There are about 20 different videos all describing the many different ways that you can create a variety of desirable residences for insects. Making one with kids is a great activity as it's a chance to talk and learn about insects at the same time and making things yourself is SO satisfying. Or if you are feeling lazy or rich, just buy a couple from a local garden centre. Less fun but quicker.

If you have outdoor lights please switch them off at night. It confuses the hell out of the poor moths. They are programmed to see lights as the moon and, as you know, they fly into them and die. If you're worried about safety, get one of those movement-activated lights. Then the garden will be dark, as it should be, but if someone does come wandering into your garden hoping to steal your bike or something, the light will go on as they pass and scare the bejeebers out of them.

As for any unnecessary concrete, tarmac or decking – get rid of it altogether. If you live in a house where someone has removed the front garden to make parking spaces for cars, dig the nonsense up and put earth there. Plant a tree – with berries for the birds. Let life happen.

And if all this isn't enough for you and you really want to get serious, you'll find a group of wonderful humans at your nearest community garden. Or how about finding a way to buy a car park and turn it into a wild flower meadow? You may need a community of people to help you, but what a joyful project that would be. New housing. For insects.

EARTH 2
Alchemy and Black Gold

I never knew that turning food scraps into rich earth felt so much like alchemy. But it really does feel a bit like a mystical art. Take pages of discarded poetry, skin of carrot, eye of potato, ear of corn, limbs of bushes, skeletons of leaves, a baby blue tit attacked – too soon – by the old tom cat, last week's cabbage, the faded roses – now not so red – a half-eaten apple, the brown-paper wrapping from yesterday's exciting parcel, tea leaves, coffee granules, last week's newspaper with journalists' carefully chosen words torn into tiny pieces, a bunch of forgotten mint from the bottom of the fridge and the yellow skins of three bananas … Place all these things in a pile in the garden. Return once or twice a week by the light of the moon to add a similar collection of black-magic ingredients and many different types of worms will appear, from who knows where, like Shakespearean harbingers of death, and, by night and by day, they will work to make gold. Or something better than gold. Beautiful, black, nutrient-rich earth. This is a black art worth a little study.

A rich soil is full of microbes. There are more microbes in a teaspoon of healthy soil than there are people on the planet. Tell that to some kids. Borrow a microscope and examine some live earth and then some dry dust (or 'dirt' as they like to call it across the pond). Soil without organic matter is dead soil. To be alive, soil needs nitrogen, phosphorus, potassium, sulphur, calcium and magnesium. All these are regularly thrown away into our kitchen bins with our rubbish.

Where I live there is a large communal garden but no one except me is much interested in magic. Years ago I asked for a hot-bin and started to throw my plant food waste into it. It felt strangely and profoundly rewarding when, later, I was able to dig earth out of the bottom and spread it on the ground. It amazes me how worms just appear from nowhere. Sometimes flies appear and then, just as quickly, vanish again. Or sometimes mushrooms spring up inside the bin and then, just as quickly, vanish again. We have a gardener on our shared patch who told me, when the mush came out smelly, that I had done it wrong. I had put in too much kitchen waste and not enough garden waste and cardboard. All those cardboard boxes I had put in the recycling – I should have torn some up and put the pieces in with the worms. But she placed the smelly pile on the bare earth and in a day or two it stopped smelling. She then mixed in lots of dead leaves and was happy. And when she thought it was all good and ready, she spread it around the garden. So for years I've been a half-hearted sorcerer's apprentice. But I didn't understand the magic nor why any of it was important.

Did you know that the earth needs feeding? That soil can be sick? That soil can be so out of balance that nothing can grow there? That soil can be dead? This has happened on some intensive farms where crops have been grown again and again in the same soil. People have gone on taking and taking until all the nutrients from the land are gone and they have given nothing back. Not even a pile of potato peel. In our gardens this happens, in our neighbours' gardens, in our parks – and we have the solution in our kitchen bins.

What's worse still is that when we throw all our organic waste and paper into bins, it can end up in landfill where, instead of doing good and helping the earth come alive, it pollutes our air instead. People think it's OK to throw out paper, cardboard and food because it's all biodegradable but that process isn't always good. On landfill sites the waste rots. As no oxygen is able to reach the rotting waste, methane is produced and methane is an even worse greenhouse gas than CO_2 (70 times worse according to some scientists). Some estimates say that a quarter of atmospheric global warming may be due to methane. So we really don't want to put it into our poor atmosphere.

This alchemy that you can learn and teach to children has a double power. First the nutrient-rich 'humus' (that's what it's called – nothing to do with chickpeas) you can spread all over the ground will enrich the soil, and so feed and nourish the plants, which will take care of the insects that will then feed the birds – and slowly the land will come alive.

The second powerful magic here is that, in a way that I don't understand, nutrient-rich earth actually captures and contains CO_2 and so keeps it out of the atmosphere where we don't want it. This process you are learning converts organic matter into stable soil carbon which plants can make use of.

Whoever you are and wherever you live, please save your raw, gluten-free, vegan waste. (Meat, dairy products, cooked food and processed foods are not suitable for this.) But please never look at a pile of vegetable peel, the discarded outer leaves of a cabbage, the skin of an avocado or a half-mouldy peach without realizing that the earth needs what you are holding. If you don't have a garden, find a friend who has one. Find a pensioner who would love both your vegetable peelings and your friendly 'hello' as you drop by to deliver them. If the elderly pensioner in your street with a garden doesn't have a compost bin, then make them one or buy them one. If you live in the middle of a concrete jungle, do some guerrilla gardening – find a patch of land that no one loves and take care of it. Or find a local allotment. Give it to a local school that has a vegetable patch. Ask a gardener you know whether they need your beautiful offerings.

Talk to the elders – those who have known how to work this magic for a generation. Have them teach you their forgotten spells. (I've even known some who add their own diluted urine to the mix.) Ask anyone who works this magic for their best tips. Learn how to be a maker of the highest-quality black-earth-gold. Trust me. It's the very best kind of magic.

THE BIGGER PICTURE 1
Being a Little Bit Activist

This is a curl-up-on-the-sofa section. Make yourself a hot drink. Are you lolling comfortably? Then I'll begin.

One friend thinks that it's a kind of arrogance to explore the difference that one person can make. According to her: 'That kind of thinking is individualism and is part of the problem.' I've always advocated that our individual actions are our greatest point of power. But according to her, if we are going to change anything, if we are going to bring about policy change at a governmental level, we have to work as a collective. The solution we agreed on is that we have to do both.

The petrochemical industries and the animal-agriculture lobby have extraordinary amounts of money and therefore influence. Decision makers are only going to make major decisions to support the climate if they are under pressure to do so. There are many people attempting to put pressure on the government in different ways. Certainly writing to your MP is one method to let him or her know your concerns. I've done this. If you're lucky you'll receive an acknowledgement of your correspondence immediately and then a month later something along the lines of:

We share your concerns but ... [cut and paste the party line on issue]. I'm not against the idea of writing to our elected representatives. I'm for it. Best to find out about your MP first, check they haven't been running a cross-parliamentary group on the subject that you're writing about, assume they are well informed, remember they are human and may have had a row with someone they love recently and, basically, be kind. But it's a slow method to try to bring about changes in our laws.

There are more radical methods. Extinction Rebellion believe in large-scale acts of civil disobedience. They say that the government won't be able to ignore the population if large numbers start to be arrested.

I'm not sure that I agree that this is the best way forward, but I thought I'd go and find out what they are like as an organization. I'm not sure what I was expecting. To be honest, I think I expected to be a little bored. I thought there would be a room half-filled with chairs, a rather earnest speaker at the front, with bad coffee and Rich Tea biscuits available. I said to a friend whom I'd invited to come with me, 'It may be boring. I've never been there before.' This isn't the way to invite a friend to an event. Unsurprisingly, she declined and I'm going alone tonight.

It's packed out. I'm on the fourth floor of an office building in an obscure part of North London on a rainy Wednesday evening and there are so many people in the room that we have trouble sitting on the floor. I look around to get a sense of the crowd. Ethnically, no diversity whatever. I'd say 90 per cent white. I wonder why this is but I think I know the answer: other communities are fighting other battles. I'd guess those present are 70 per cent under 40 and possibly 60 per cent female. But there looks to be a good class and career mix: from the heavily tattooed and proud hippy look to the expensively dressed professional.

The man organizing the evening has a friendly but frantic expression. He says there is more to get through than can reasonably be achieved in one evening. He introduces himself: 'I'm one of the activists who stripped naked in the House of Commons demanding climate action.'

'Could you show us what that action looked like?' I ask. The room laughs as he takes off the top two layers and then disappointingly stops. But there's little time for humour – or stripping.

He gives us a quick introduction to the 'organization'. I use this word with some hesitation as they're not really an organization – more a group of people who care about the planet coming together. But they do have an impressively long list of rules. He hands out the leaflets and we read them out loud so they can be sure we know them. There will be no one turning up with a beer on any Extinction Rebellion events.

1. 'We will be strictly non-violent in our actions and communications with members of the public, workers, the authorities and each other at all times.'
2. 'We will treat everyone with dignity and respect.'
3. 'We recognize that non-violence is essential to our campaign, whilst recognizing that using non-violence is a privilege that is not available to everybody.'
4. 'We will act calmly and carefully and will strive not to endanger people.'

5. 'We will not take action under the influence of alcohol or drugs.'
6. 'We will keep ourselves informed of the legal consequences of our actions and take responsibility for them.'
7. 'We will honour our regenerative culture, looking after ourselves and each other in order to take effective action. Safety will be a high priority at all times.'
8. 'We will plan actions with care, being mindful of blockades that may affect emergency routes.'
9. 'We will consult our traffic plan to mitigate for these eventualities.'
10. 'We will not take action unless trained and aware of the consequences of our actions.'
11. 'Alongside protest and civil disobedience actions, we encourage constructive direct actions that offer solutions to the climate and biodiversity crises we face.'
12. 'We encourage the creation of a regenerative culture, that supports us as we focus on these solutions in more connected, cooperative communities.'
13. 'We will tell stories of positive change to encourage others to act. This will build a stronger narrative that complements our opposition to failed business-as-usual approaches.'
14. 'As activists in the Global North, we collectively acknowledge our privilege. We are on an ongoing journey of understanding what this means. We act in solidarity with social and environmental justice frontlines outside the UK and recognize that our struggles are connected.'

This is impressive.

Someone speaks about what's happening to climate change activists in other parts of the world. It makes me grateful to be in Britain.

We have a few quick introduction exercises: break into small groups, say how you're feeling and why you're here. Then the organizer is immediately on to forming 'affinity groups' also known as 'finding people like you'.

He smiles at us.

'If you are willing to be a naked climate activist – stand there. If you are willing to wear very little – stand here. If you are willing to make and wear extraordinary costumes and walk the streets in them – stand there. If you are only willing to go out in the street in your normal day clothes – stand there.'

We each shuffle through the overcrowded room into little groups.

'Now see who you are standing next to and introduce yourself.'

We've barely had time to do this when he repeats the exercise.

'If you have particular skills and would like to use those skills to help the planet, we need to match you with others that have those skills. For example – if you are a doctor, a nurse or have medical or first-aid training – stand there. If you are an artist and able to make things or paint things – stand there. If you are media savvy – good with photos or videos – stand there. If you are a musician – stand there. If you like working with children – stand there. If you have legal experience – we need legal observers; you will have special jackets that state clearly you are a legal observer and the police are very careful when you're around – stand there. If you like to look after others or are a qualified counsellor – stand there. Now introduce yourself to those around you.'

A couple of the same faces keep appearing beside me. One woman I meet is a climate scientist. Another is a Pilates teacher. One man introduces himself as a 'grandfather'.

The organizer continues: 'If you definitely want to be arrested for this cause and you have decided that a mass uprising is the only way forward – stand here. You will receive special training.'

'If you will be a little bit disappointed if you don't get arrested – stand there.'

'If you're not sure whether you're willing to be arrested or not – stand there.'

'If you definitely don't want to be arrested – stand there.'

This is challenging. I usually like to be on the front line but I want to be a climate activist not a criminal. On the other hand, women only got the vote because the brave ones were prepared to be break the law. What are the legal implications of arrest? I'm unsure. So I join the 'unsure' group.

Around those of us listening to this talk, other people are working. A group of artists appears to be making a dinosaur. Someone keeps sawing wood; he's making a compost toilet to install in the middle of Marble Arch.

There are way too many people here. It's a friendly, slightly scared, excited frenzy. Bored? We're not.

'Hello,' I say to a Pilates teacher who I'm meeting for the third time.

'Now call over any people you want, to make groups of about eight people. This is your "affinity group."'

I look for our climate-scientist friend, but she's joined the 'I'm willing to be arrested' group.

We have lists of different activities we can choose to support. I'm encouraged by the set-up. No one's being asked to do anything that they don't want to do.

'The most important thing is that everyone should be happy and comfortable with what they are doing,' the formerly naked man explains.

A woman from the legal team introduces herself: 'If anyone would like to know the legal implications of being arrested – see me.' And she vanishes again.

'I'm Zoë,' says the Pilates teacher. She asks, 'There are two weeks of climate change action starting in three days. So what does everyone want to do?'

'Can I just explain that I only came along tonight to find out what was happening. I was expecting a draughty church hall with a group of elderly hippies. Not a Central London venue overcrowded with young and determined activists,' I contribute unhelpfully.

'OK, so I hope it's a pleasant surprise. Now there's going to be a camp in Hyde Park. I'm a qualified first-aider so I'm going to be in the first-aid tent.'

'Isn't it illegal to camp in a Royal Park?' asks someone.

'Yes. But what are the police going to do with thousands of people with tents?'

'Intense "in-tents" non-violent hippies with a purpose? I'm not sure they'll know what to do.'

What have I walked into?

'I'm going to join the artists. I love making costumes,' says one woman and leaves our group to find her fellows.

They give out information sheets with the intended actions.

The once-naked speaker demonstrates hand signals for achieving silence. They work perfectly. 'Any questions?'

A bearded man asks, 'This blocking the road. I'm concerned about the emergency services. I don't want anyone to die because an ambulance can't get through. That's not responsible.'

Everyone applauds.

The speaker explains. 'The emergency services know in advance exactly which roads we are blocking and how to get around them. We have to co-ordinate with the police to a certain extent. That doesn't mean we have permission. But they know what we're doing and that we're all non-violent.'

Another round of applause.

Someone stands up and gives a short unsolicited speech about climate change. This is what you call 'preaching to the choir'.

We listen patiently.

Someone else stands up and says that this will achieve nothing apart from alienating Londoners from the cause. What will annoying drivers actually achieve?

I have my own reservations. Mainly that the *Guardian* will write about the importance of the issues being raised and every other newspaper will describe everyone as 'troublemakers and hooligans'. With a neighbour who reads the *Daily Telegraph,* I'm usually able to see the world partly through his eyes. Sadly, large parts of the country see everything in this way too. I keep quiet.

Random voices shout out questions.

'Don't you think blocking traffic is irresponsible?'

'No, we think it is necessary. We think that the government ignoring climate change is irresponsible.'

'Isn't there a better way to move climate change up the political agenda?'

'If you have ideas our office would very much like to hear them.'

'Do you have people trained to speak to the media?'

'Yes, we do.'

'Do you have many people that are willing to be arrested?'

'Yes, we do.'

'Do you have people who can give legal advice?'

'Yes, we do.'

'Can we go to the pub now?'

'Yes, we can.'

That was an extraordinary evening.

'What just happened?' I ask Zoë.

'You just signed up to be part of a revolution.'

'Ah.'

So now I'm home. It's the evening and I have time to think about this.

To be honest, I'm a bit freaked out. Planting native species in the garden to attract insects is one thing. Getting a criminal record is another. And I did have a guideline that I was to do all this joyfully. I look through the list of ways that I can contribute – and discover they have a samba band.

I play traditional Japanese taiko drums. It's an ancient form of drumming that is as much to do with the visual form as it is to do with rhythm. Surely being bad at drumming must be a transferable skill? There are two days of teaching before the action begins. Then they will let this newly formed samba band loose to support morale. To march up and down the blocked streets. To make music – or at any rate a rousing noise. This sounds huge fun, is relatively harmless but still fairly frontline. I'm there.

*

Before that, a treat – because today is a Friday. Friday is the home of Fridays for Future, the international movement started by Greta Thunberg. She missed school on Fridays to raise awareness of her government not having strong enough policies to reverse climate change and ecological breakdown. The argument is simple enough: what is the point of an education if governments won't listen to 97 per cent of educated scientists? Why learn science if it has no influence on policy?

What previous generations have done is immoral. They have gone on creating this climate disaster thinking that 'future generations' will find a way to undo the damage. Presumably future generations are also supposed to come up with ways to reverse biodiversity loss, reintroduce species that have become extinct, bring our soils back to life, clean up the air, get the plastic and the acid out of the seas and put coal back in the ground. The legacy we leave our children and grandchildren has become impossible to reverse.

Many of our highly intelligent young people understand what has been done. There has always been an unspoken contract between generations: that we leave a better world for our children. And we've broken it. They're angry. And they have reason to be. So tens of thousands of them are on strike. All over the world. Every single Friday.

'I've never heard about this,' says a friend who is a teaching assistant at the school where members of the UK Royal Family send their children. Well, no – the tragedy is that children from the most wealthy families (who may also be passionate about the environment) don't hear about this because many expensive private schools want to uphold the status quo and would never allow their students to be involved in any matter as insignificant as trying to force our government to have a strategic policy to reverse climate change.

But in many cities around the world, representatives of the dumped-upon next generation are out of school whether their teachers like it or not. So I cycle down to Parliament Square to check out the youth strike. What an astounding blast of passion they are. I weave my way through the hand-painted banners. Some are practical: 'Don't be part of the pollution, be part of the solution', 'Climate change is bad for the economy', 'Leave the coal in the ground', and the now famous 'There is no planet B'. Some are very direct: 'Let's fuck each other, not the planet' and 'Keep the Earth clean – it's not Uranus'. There's even a bit of street slang: 'Real ass bitch give a fuck bout the planet'. I love 'Spill tea not oil', 'Frack off gassholes' and 'I am the Lorax. I speak for the trees – they say "fuck off".'

They move on to the 'shouts':
'What do we want?'
'Climate justice.'
'When do we want it?'
'Now.'
There's also:
'Whose future?'
'Our future.'

And, in case the police forget:
'Show me what democracy looks like.'
'This is what democracy looks like.'

I just stand with the tourists taking pictures. There's also a lot of filming going on. One young girl (aged 12 – I asked her) introduces herself as the children's climate change activist for education at the United Nations (UN).

She is asking anyone she can find to interview, 'What are the main causes of climate change?' And the young people answer her in detail. I watch the interviews. These young people understand climate change and what needs to be done better than most adults. The youth here went past being 'a little bit activist' when they were ten. They have now reached 'experienced activist'.

They shout, they chant, and then they march from Parliament Square to Trafalgar Square. A girl runs to catch up the group. The banner she's carrying reads: 'Make love not CO_2'. Yup, she has all the solutions to human happiness and planetary survival right there in four words.

On Saturday morning the rehearsals start for the samba band. The meeting place is the middle of Regent's Park, and when I arrive for session two, it seems that that even session one has been enough for the tender sensibilities of the Regent's Park residents. The police have asked the drummers to move on.

A solitary figure, Jake from Somerset, looking mega-hippy with a van full of drums, is posting the change of location on Facebook.

'I just teach samba drumming,' he smiles. 'I'm not used to this level of stress.'

The friendly police suggest Russell Square. Obediently, we relocate.

People appear from various directions. We have 30 drummers with small tom-tom drums (I choose one of those), larger tom-tom drums, base drums, cowbells and various percussion instruments.

Drumming is absolutely joyful. It just is. There's a line in the Psalms that I know (from a previous lifetime in this life): 'Make a joyful noise unto the Lord.' We are making a joyful noise to Mother Earth. I hope she appreciates it. The residents of Russell Square are enjoying it: it's a sunny afternoon with free entertainment.

The different rhythms are learned by remembering words. Some of the rhythms are surprisingly straightforward. Base drums: 'We like Potatoes.' Boom Boom Boom-Boom-Boom. Tom-toms: 'Do you?' Bam Bam. Base drums: 'I like them mashed.' Boom-Boom-Boom Boom.

Tom-toms: 'I do.' Bam Bam.

And another that works as a call and response.

Leader: 'We all want some bun-ny ears.' Bam Bam Bam Bam Bam-Bam-Bam.

Whole band: 'We all want some bun-ny ears.' Bam Bam Bam Bam Bam-Bam-Bam.

Leader: 'What do we want?' Bam Bam Bam Bam.

Band: 'Bunny ears.' Bam Bam Bam.

Leader: 'Bunny?'

Band: 'Bunny.'

Leader: 'Bunny?'

Band: 'Bunny.'

Leader: 'We like.' Bam Bam.

Band: 'Stroking our bunny ears.' Bam-Bam Bam Bam-Bam-Bam.

I smile while learning this rhythm, thinking that climate activists are seen as troublemakers or are even compared to terrorists – and here we are indoctrinating our resistant brains with propaganda such as, 'We like stroking our bunny ears.'

During the breaks people ask us who we are and what we are doing. We explain again and again.

'Good for you,' says one older man. 'I'm one of the founders of the Green Party in Germany. I appreciate your passion for the planet very much.'

'One of our drummers is German too.'

I point out the founding member of the German Green Party to the drummer. 'Would you like to say hello?' I ask him.

'No thanks,' says the drummer. 'The German Green Party is so far up the arse of the right you can't tell the difference.'

'Ah.' Perhaps I won't mention this to the founder appreciatively smiling in the sunshine.

We learn four or five pieces, as well as the breaks and how to change flawlessly from one piece to another.

We play all afternoon until an elderly lady complains that we are disturbing the peace. She isn't wrong. We are. So, like hard British anarchists, we apologize profusely for upsetting her, stop immediately, promise that we won't play here again tomorrow and wish her a pleasant day.

'Where is a big patch of common ground that's reasonably near London?' asks Jake.

'Hampstead Heath?' someone suggests.

'OK. Tomorrow morning. 9am. Hampstead Heath Station.'

9am? On a Sunday? Sheesh, these activists really are committed.

I arrive at Hampstead Heath Station to find a group of early risers already unloading a van of drums and walking onto the Heath. We meet a park official who says we can't practise there.

'They complain about everything here.' She's friendly.

'Could you arrange for there to be a small problem with the batteries on your phone this morning?' I suggest. 'We're practising to support environmental activists this week.'

'I'll walk in this direction.' She leaves.

A surprising number of people arrive. It's a beautiful sunny morning.

'Isn't it stunning here?' I say to one older lady who is clutching a cowbell she's learning to play.

'It depends on what you look at and what you notice is missing,' she says.

'What do you mean? What's missing?'

'The birds. Look around you. There are no birds.'

It's true. I can't see a single bird. On Hampstead Heath, which is a protected area.

'When I was a child the skies were full of birds and the sound of birdsong.'

I hadn't noticed. There isn't the sound of a bird anywhere. One solitary seagull flies in the distance. 'Look – there's a pigeon.' I remain determinedly cheerful. 'So that's why you're here?'

'Yes. I'm here for the birds.'

Back home there are plenty of birds singing in my garden. But we garden organically, compost, feed the earth and look after the insects so the birds have something to eat. This is a desert.

'If we play here we won't be overlooked,' says Jake. 'It may hold off the complaints.'

This is a green and pleasant place to practise.

And so we start:
'What do we want?'
'Bunny ears.'
Radical troublemakers.

We drum for two hours. People are out walking with their families and their dogs. A crowd gathers. Children dance. It's a primal thing with drums: everybody seems to love them. Almost everybody.

A police van drives across the park and pulls up at a distance. They watch the people dancing.

Jake looks nervous. 'Oh shit, I'm just trying to teach some drumming here.'

I stroll up to the van. 'Good morning, gentlemen. Lovely day, isn't it?'

'It is. Great drumming.'

'You're not insisting we stop?'

'You notice that we've not got out of the van,' one says. 'People complain about everything that makes noise here.'

'Except the birds.'

'Sorry?'

'We're environmentalists. Someone pointed out there are no birds left on the Heath.'

'There are swans on the lake.'

'Yes. And a few pigeons and seagulls, but as an example of catastrophic biodiversity loss, Hampstead Health, for all the wealth of the residents, is a pretty good one.'

'About those residents.' He sees an opening while the drumming goes on in the background. 'There are 47 bye-laws for things you can't do on the Heath. You can't hold public meetings for prayer, have poetry readings, sing, play music, collect money for charity, play sports, sail model boats, beat carpets or wash your dog. Dancing is a complete no-no.'

'Or having sex?'

'Well, there is a historical precedent about having sex. But you shouldn't be seen. That's part of the fun apparently.'

I laugh. 'So having sex you can turn a blind eye to … but drums?'

'Hard to turn a deaf ear to.'

'Ah.'

'We have mixed feelings ourselves. The vast majority of the public is loving the free entertainment. We can see that. And some of the residents here are, well, not the nicest people. One man just stopped us, pointed at you drummers and said, "Clear that scum off the Heath immediately." We don't like that sort of instruction. So we drove slowly. Could you be so kind as to tell us when you plan to stop? Maybe another 30 minutes?'

'We may have a problem with that.'

'Do you have permission to be here?'

'I don't think so.'

'In that case we'll have to ask you to move on.'

'Oh dear. OK, I'll go and speak to our teacher.'

'Who is in charge?'

'No one. We have a music teacher but he's just a lad who teaches drumming and he's super stressed and not used to dealing with intimidating authority figures like yourselves.' They were the least intimidating men ever. 'What law are we breaking anyway?'

'It's a bye-law. Playing an instrument. A guitar or a flute is also an offence.'

'I understand. You gentlemen have your jobs to do.'

I stroll slowly back to the drummers. 'Jake – they're asking if we can move on.'

'No, we can't. I'm just sending everyone off for lunch.'

I stroll back to my friends in uniform.

'There will be no noise for the next hour but we need at least one more lesson this afternoon for those just arriving.'

'Can you keep it to an hour?'

'We'll take that.'

And they drive away.

So we welcome new drummers and we sing our parts. At 3pm, four new park police arrive to enjoy the drumming.

When we start to play, a new crowd forms, cheering us on, especially when one angry male member of the public shouts at Jake. The police step in to take him aside so that we can continue. This is an interesting reversal. The police protecting us. We play on for an hour. Then stop. Someone asks, 'You happy with that?' And I believe the police may have said that one more session wouldn't hurt. Or that may not have happened – I'm not sure. All I know is that we play on, the crowd dances, the sun shines and it's a truly joyful afternoon.

Monday morning is different. I'd be lying if I said I'm not nervous. Alarmed messages come in from friends who can see what I'm doing through social media. 'Isabel, the consequences of arrest are becoming increasingly serious. Don't do this.' 'If you're arrested, don't resist – that makes it worse.' 'Please paint your face – it confuses the face recognition cameras.' 'Please wear an animal mask or something.'

One of the rules of Extinction Rebellion is that you don't wear masks. You come as yourself or you don't come at all. But as a concession to the friend that wants me to paint my face, I draw on a heart with eyeliner and lipstick. I find a hat and set off to Waterloo Bridge, which Extinction Rebels are planning to turn into a garden with trees, grass and flowers. Rather ironic as some Londoners have been asking for a 'garden bridge' for years. The activists are hoping to give us one.

It's hot. I'd chosen to wear a smart woollen coat.

On the train a man I'm sitting next to smiles at me struggling with a drum, two drumsticks, a banana, a Danish pastry and a coffee.

'Can I hold something for you?' he smiles.

I laugh, 'I do seem to have been a little ambitious with how much I can carry. Thank you. But when I've eaten the banana it will all work.'

'You've smudged your heart.' He looks at my cheek.

'Ah well,' I say, 'It's already broken, bruised and battered. A little smudge can't hurt it any more.'

'I hope you're speaking metaphorically?'

'I suppose so.'

'They heal apparently,' he says.

'Do they?'

'I fucking hope so.'

'I don't wish to discourage your optimism but mine has been broken for about ten years and isn't showing any signs of repairing itself.'

'Mine neither,' he sighs.

'But maybe yours will be faster. You're obviously saner than I am.'

'Why do you say that?'

'Well, look at me. Sat here with a smudged lipstick heart on my face and a drum.'

'It's the other way around. Look at the state of the world. It's the person with the drum who is the sane one.'

What a lovely thing to say.

'That's very sweet of you.'

In a movie this would signal the start of an exquisite romance. But I write real life. He smiles, waves and gets off the train. You may think that I made this conversation up. But no – that was it, word for word. I feel encouraged getting that vote of confidence from a stranger.

Outside Waterloo station we drummers find each other. There is a delay as Waterloo Bridge has not yet been 'taken'.

I love that they say 'taken' rather than 'blocked': it makes me feel as if I'm in the middle of the storming of the Bastille. There is a buzz of nervousness and excitement.

'How are they going to turn the bridge into a garden?' I ask.

'They have vans full of trees in pots, flowers, plants. Strength in numbers, I guess.'

I introduce myself to strangers who have travelled from Bristol. Jake looks exhausted and we haven't even started yet.

Then someone shouts, 'They have taken the bridge! The bridge is secure!' Heaven knows what the police would be making of this. Hippies with trees in the middle of the Monday-morning traffic.

Jake blows his whistle signal. A crowd gathers. We must be between 25 and 30 drummers and an interesting mixture of percussion instruments from cowbells to tambourines. The base drums really make a really powerful sound. And we know some good rhythms now. Joyful? Absolutely. We march off to the delight of the morning commuters. Tourists hold up cameras. We watch Jake's hand signals: crossed arms for a change of tune. You don't want to be the only one playing when he has just counted in a silent break. One drummer missing the silence ruins the magic of the drums stopping and coming in again all as one.

Then we come across an extraordinary sight: a row of police standing in front of Waterloo Bridge with a huge banner behind them that says 'Climate Emergency'. On the bridge, trees are being secured down the middle and people are putting up tents. Children are drawing flowers on the pavement with coloured chalks. We drum our way past the policemen and up onto the bridge.

'Hold on a second,' I think to myself, 'Haven't we just saved the police the trouble of kettling us by kettling ourselves?

Who was blocking the bridge the other side – activists or police? Anyway, we perform a couple of pieces and then break for tea, which is being given out in the sunshine.

Workshop spaces are being set up. One group is erecting a yurt and goodness knows how many trees are appearing. I'm glad that I'm not the police professional in charge of trying to work out how to get the bridge back. The options are clear: either send in the tear gas and tasers and shoot people as they would in some countries, or let the people express themselves. There are chants from the people for the police, 'We are non-violent. How about you?'

Families with young children spread out rugs for picnics. Someone juggles six clubs skilfully. Journalists and camera operators walk up and down wondering who to talk to.

Then we're up and marching on the bridge again. I wonder briefly whether this is an example of drumming to the converted but we're entertaining both climate activists and the police. Then they ask us to stop. A group of sound engineers have put together a 'Requiem for Birdsong'. They have a speaker and, as we stand silently, they play birdsong. This is the sound of the summers of our childhoods, a beautiful dawn chorus that would be completely unknown to many young people who live in cities. And – as everyone who listens knows – it's a sound that will become history unless we all change our ways.

Tears prick the backs of my eyes. It's such a beautiful sound that most of us have taken for granted. We've assumed that this sound will go on forever. It may not. Each time we buy food that is not organic, we are supporting the use of pesticides. This kills the insects so the birds have nothing to eat. We have parks that are full of nothing but grass and permitted trees that don't bear fruit. I resolve, as I listen, to learn more about protecting the birds in my own area. And to appreciate them more. I can only recognize the song of a robin and a blackbird, the screech of the green parakeet and the coos of the pigeons. The recording ends. No more birdsong.

Someone signals to the drums. Jake blows his whistle and we march straight off Waterloo Bridge. 'We've been asked to go to Marble Arch.' So we all get on the Underground.

There are actions in different locations. At Oxford Circus, we hear that a pink boat has appeared in the middle of the major junction with Regent Street with 'Tell the Truth' emblazoned on its side. This is an instruction, not to the climate activists to be honest when being arrested, but to the government and the media. The activists are asking them to be honest about climate change.

At the Shell Centre, one of the world's most famous international environmental lawyers, Farhana Yamin, who played a leading role in negotiating the Paris Climate Agreement, is one of those who has glued her hands to the building to highlight the culpability of the company.

I'm a Londoner. I've lived here all my life. But I've never come out of the Underground at Marble Arch to find no traffic and a party going on. It might sound a ridiculously obvious thing to say, but cities are wonderful without traffic. The air pollution levels, normally toxic in Zone One, are down; people walk while bicycles speed by. It's a positive vision of the future. Except it's all illegal. Jake blows his whistle and we put on our drums and play.

It feels good to be part of a samba band. People's faces light up as they listen to us. Children immediately start dancing. Then Jake calls silence and we carry our drums silently across the park to the place where the camp will be set up. From every direction people are arriving with tents. Some have been on a long symbolic walk to raise awareness of the climate crisis, banners read 'Totnes', 'Essex' and 'Land's End'. How their feet must hurt.

The activists gather in a huge circle to listen to speeches. People sing and read poetry. Then a policeman takes the microphone. 'I'm a Druid,' he says.

A cheer goes up. Ha ha. Of course he's not a Druid. He's using what must be the first rule of good policing: establish that you are not the enemy.

He continues, 'As you know, this is a Royal Park so you are not allowed to camp here. But we realize that you people are non-violent and have no drugs or alcohol. We know why you're here and we care about the planet too. So on this occasion we're allowing you here. You're the most polite collection of people I've ever met,' says the Druid policeman.

We cheer.

Families with tents pour in from all sides of the park. A Hare Krishna group arrives and hands out free dahl and rice. There is water from a tap with cups made from oat-milk cartons.

'Where next for the band?' someone asks.

'We have a call to play at Piccadilly Circus which is held by the youth groups.'

We strike up and march down Oxford Street to get to Piccadilly Circus. 'What do we want?' 'Bunny Ears,' ha ha ha. When we reach Piccadilly, Eros, the God of Love, is waving several Extinction Rebellion flags. Around the base of the statue it now says 'Climate Emergency'. A young man climbs up onto the statue.

'Youth, students, climate activists – welcome to Piccadilly Circus or, as I prefer to call it, our new home!' Cheers and laughter. The police are smiling – I think it's hard for them not to. They could remove the young people from Eros and arrest them for 'damage to a public monument' but they have enough to do without arresting environmentally conscious young people.

Our band has one banner that says 'Climate Emergency'. The youth groups have much better banners: 'Listen to our Peer Review Scientists' says one. Another is clear on what the climate activists want: 'Zero Carbon 2025. Carbon Negative 2030'.

Some are clear in other ways: 'Where the Fuck is the Government?'

I also love the signs that say: 'We Value Reflecting and Learning' and 'Humility, Empathy, Frugality'. A young girl gives out cloth patches. She offers me one that says 'Conscientious Protector'.

'Yes, please.' I love that. I pin it on the shoulder of my woollen coat. The sun is as hot as midsummer although it's only early spring. Some men abandon their tops all together. It's a demonstration and a party.

Jake blows his whistle. We stand up and do a 'set'. The rhythms raise the spirits and pump up the blood. Everyone dances. As I play I'm half aware of the miracle that's taking place around me. In the face of generations of political failure from successive governments, which have become beholden to the money and power of the petrochemical industry and the animal agricultural industries, here are young people using the only power that they have to save their future: the power of numbers. And they have shut down Piccadilly Circus.

I might have felt sorry for the shopkeepers. But the shops here are not small independent shops run by people who care about what they are selling. They are exploiters of both planet and people. If Primark sells fewer clothes for a week, I don't care. Anyway, the presence of young people dancing is hardly intimidating. Anyone who can walk or bike can shop. It's only fossil fuel driven transport that can't get through.

We finish drumming. I see a woman in tears standing alone in the crowd and walk up to her. 'Are you OK?' I ask stupidly.

'I am. It's just that this is so important and my love of the planet is so profound, I find it hard to put into words.'

'Yes.'

'These young people. We – the older generation – we knew that what we were doing was immoral. Creating atmospheric pollution, killing the oceans. We knew and we just said, "Someone in the future will find a solution." But our children, our grandchildren … they have the right to an unpolluted planet without the diversity loss. They have a right to have children that can see animals and birds in the wild and …' She breaks down into sobs.

'I know. I know.'

'Why should it take these young people to shift the immoral greed of the previous generations? The lack of care? The bad stewardship of the planet?'

'I know.'

'And they call these young people criminals and troublemakers? It's our politicians and the greed of the super-rich who are utterly morally corrupt. They are the ones that should be arrested.'

'There are a lot of people who would agree.'

'I, I …' She sobs. 'I just don't know what to do!'

'I have a suggestion,' I reply, Tigger-like. 'You could do what I'm doing.'

'I can't play drums.' She looks at my drum.
'No, I don't mean that. Sit down a moment.'
'On the ground?'
'Why not?'

Jake and the other drummers are sitting down, chatting and eating. The blessed Hare Krishna group are giving out free vegan food again. How I love them. They believe that all food must be cooked with love – that added invisible ingredient.

My new friend blows her nose.

I give my speech: 'What I'm doing is this. I'm looking for every single way that I can to help the planet. It touches everything: how we work, travel, shop, food, bank, dress – everything. You could do what I'm doing.'

'Well, I recycle and I don't take plastic bags.'

'That's a good start. But more and more people are questioning whether recycling is really as effective as we would like to believe. There's a movement to become totally waste free. So you don't just recycle – you find ways not to take packaging at all. It's not easy but it's fun.'

'Fun?' she asks. 'It sounds impossible.'

'Fun is a vital part of it. Because life is short. So, if you like, you could do not only what I'm doing but the way that I'm doing it. I'm doing all the stuff that spreads joy to myself and others.'

'That's why you're drumming?'

'Correct. I wasn't sure if I wanted to get involved with Extinction Rebellion. I'm not sure I agree with blocking roads. They've taken care to inform the emergency services and if we hear an ambulance everyone clears the roads in seconds. But if I knew that just one person had died because an ambulance couldn't get through, it would question everything that's being achieved.'

'I do agree.'

'On the other hand, as they say, they have resorted to disruption because peaceful demonstrations and petitions have been ignored for years.'

'I'm really not an activist. I'm more of a gardener.'

'Then be a gardening activist. Did you know that, to maximize the chances for our insects, we need to plant 70 per cent native UK plants in our gardens?'

'I didn't know that.'

'So, that's my point. If you're a gardener you'll probably need to change the way you garden. Everything needs to be considered. Do you eat meat?'

'No.'

'That helps. Then there is the impact of our eating, of course – locally grown, in season, all that. Everything is impacted. No more domestic flights. All flying has to be questioned. Your shopping will be influenced. As it says on that girl's armband: "Frugality".'

'I see. Yes.'

'But it has to be joyful.'

She smiles. 'Because life is short.'

'You got it.'

'Thank you. My name's Marion. You've been most helpful.'

'Good. It's a beautiful day. Something remarkable is being achieved here. Let's find positive ways forward.'

'Do you think it's all right to eat fish?'

'Personally – no fish for me. I like my fish alive, in the sea and breeding freely. Why do you ask?'

'It's just that I've been feeling this. I think I'm ready now to stop eating fish.'

'The poor creatures are full of microplastics anyway. You'll be doing your health a favour.'

'Yes. Thank you so much.'

And off she went. Well, that was quite a conversation.

Jake wanders over. 'You done for today?'

'Yes, I think so. I'll see you tomorrow.' And I descend into the Underground, proudly displaying my drum with the Extinction Rebellion logo painted on the side in blue and green.

'Thank you for what you are doing,' says a man on the train.

But that night I lie in bed tossing and turning. Am I being an insipid coward? If I had been alive when women were fighting for the vote, would I have been one of those brave women chaining themselves to railings? Would I have been in prison on hunger strike? Would I have been one of the numbers that really made a difference, or would I have just been banging a drum? If I had been in South Africa when so many, both black and white South Africans, were showing astounding courage to try and shift the status quo, would I have been drinking tea and ignoring the arrests?

I admire the 'arrestables' who have decided that they don't mind being arrested in order to prove how many people care passionately about our planet. Why am I not amongst them? Should I be gluing my hands to the Shell Centre? Do I really imagine I'm helping the planet by banging a drum? No answers come back from the universe. With bunny rhythms in my ears, I fall asleep.

I wake in the middle of the night from a dream in which I'm being told that a tidal wave is coming and to hold on to something. I'm not holding on, I'm looking around for my mobile phone, which just about makes me a perfect parody of our age of insanity.

*

The following morning, as I sip my tea rather groggily, a neighbour arrives at my door. He's a professional pianist and he's dressed up in his black coat-tails at an unusually early hour of the day for him.

'Got an early gig today? Going somewhere nice?'

'Very much so. Marble Arch.'

'Really? You're going to the climate action?'

He hands me a leaflet.

Over 40 London-based professional pianists from around the world will gather to perform a groundbreaking ceremonial concert dedicated to the world's climate refugees. They will add to the growing international chorus calling on governments to respond adequately to our 'direct existential threat'. (Antonio Gueterres, Secretary-General of the UN)

'Goodness.'

'They have an amplification system for the piano which is being run by pedal power so the piano has groups of cyclists on either side so that everyone can hear the music.'

'I love this.'

'A lot of my professional pianist friends are involved. We've all got to do our bit, haven't we? I thought you'd be pleased.'

'How did you know I was involved?'

'I saw your FB post.'

The BBC had run a picture of the band. I'd posted it with pride.

'I was only a dot in a hat. But there I was in the third row of the band. I guess every dot counts.'

'At the storming of Waterloo Bridge?'

'Exactly.'

'*Vive la Résistance!*'

'Or something. Yes.'

'Well, I must dash to get to my slot.'
Amazing. Even my next-door neighbour is involved.

My shoulders hurt and my thighs are bruised from the tom-tom drum banging against my legs. I stick plasters over blisters on my fingers, wash my hair and floss my teeth for an extra-long time to give myself the illusion that something is under control. I check the band's WhatsApp message thread, take my drum and go off to play. And the days go by: 'five, six, seven, eight'.

Every time someone is arrested the crowd cheers and chants 'We love you.' This confuses the police who are trained to believe that being arrested is a bad thing. I watch people either being walked away if they don't resist arrest, or being carried if they do. So much courage. I feel a total wimp.

One evening we play at a road block for over an hour. The occasional policeman wanders over, but they have obviously decided the band is just pure fun. Oxford Street is so much better with just pedestrians and pedal bikes. The people have 'reclaimed the streets'. Surely with most of Central London taken over with the aim of saving our planet, surely our government has to listen?

On the other side of my house to where the pianist lives, is my neighbour David, who reads the *Daily Telegraph*. I ring and ask him if he would kindly drop by anything he sees about the international climate action..

The following morning he brings over the newspaper's leader column. It is often written by the editor or the deputy editor and, interestingly, is usually unsigned. It's titled, 'The Dark Side of Pink Protesters'. How strange – why call us 'pink protesters'? Is it because the boat is pink and he's seen a picture of it? Or has he chosen 'pink' because he doesn't want to acknowledge that Extinction Rebellion are green protestors? Or is it because many Conservatives are passionate about the environment too and he doesn't want to highlight any positive pointers before they've started reading?

By the end of the first sentence he has stated that 'protestors have vandalized dozens of buildings'. This just isn't true. As far as I'm aware, one building has been vandalized in a symbolic act – the Shell Centre. Of course the writer doesn't mention that it was the Shell Centre. I'm very disappointed by the *Telegraph*. Call me old-fashioned but I do expect journalists to have a passing regard for the truth. Extinction Rebellion says that the media doesn't tell the truth and here is the evidence in the first sentence. He goes on to say that the protestors have set out to cause 'wanton damage'.

In the second sentence he says that the activists are part of an 'allegedly' global movement. How can he get away with this? Five minutes' research would let him know that there are environmental activists using the same non-violent principles and working under the banner of Extinction Rebellion in 80 cities in 30 countries.

The journalist's use of language is laughable. He says that the climate activists belong to an 'outfit' called Extinction Rebellion. He goes on to explain that some claim a 'kindred spirit' with Greenham Common women – and then lets us know why the Greenham Common women were in error. He states that the activists are doing the worst thing known to humanity: making demands of the British Government which would put the UK at a competitive economic disadvantage. He doesn't mention the economic advantages of a sustainable green-energy policy.

I go on reading: 'The sanctimonious climate change activists essentially want to put us back into the dark ages by ending economic growth. They are inflicting misery on commuters and businesses for no purpose whatsoever other than to claim a spurious moral high ground.'

Readers who have studied the remarkable work of the spiritual teacher Byron Katie can allow themselves a smile at this one. Katie has a method for showing that the exact thing we accuse others of can often be true of ourselves – and rarely have I seen a better example of this. How much more sanctimonious could the writer be? And as for a moral high ground, I'd argue that those who are locked under boats and bridges trying to save the rest of us from catastrophic climate breakdown don't have to 'claim a moral high ground' – they already have it.

How is this man (at the time of reading, both the editor and sub-editor are men) allowed to put his factual errors, ill-informed opinions and insidious slurs in the leader column of a leading national newspaper? I have a quick look at other coverage. The *Independent* has an article headed 'Radical Action has never been more justified'. One article on the BBC website says climate protests are 'diverting' London police. I have to smile. The London police have certainly been diverted in more ways than one. I'm not sure how the use of inverted commas on either side of the word 'diverting' makes it clear that it means 'have taken police away from more important tasks' rather than 'entertained them with singing, drumming and dancing'. I scroll down the online BBC article and there at the bottom is a link, 'What is Climate Change?' And there it is. Just in the one link: the evidence that the actions are working. They have actually written an article, below which a link has been provided that indirectly says, 'Climate change is real – learn about it here'.

That link alone was worth drumming for. I glance online at the title of a *Guardian* article: 'The Guardian's View on Climate Change Campaigners: Suited or Superglued, We Need Them All'. It includes the story of an 80-year-old woman who had locked herself to the bottom of a lorry, saying that she refuses to leave 'a broken and barren world for her beautiful grandchildren'.

I open the *Sun* newspaper and discover that I am part of a 'far-left mob'. Ah yes, that old chestnut. I'm sure my Oxford-educated, Conservative-voting climate-change friend from last week's event would be interested to know that she's part of a 'mob'. One of the labels Extinction Rebellion hand out that I have seen attached to drums and jackets reads 'Beyond Politics'. If only they would let us be. It's tempting to spend the morning watching the media reaction, but today the call is to take a drum to Marble Arch.

The sun is warm when I arrive and there is a joyful, festival atmosphere. Someone has bought extra drums so we'll play and invite anyone listening who's a drummer to join us. We play for a while and in a short time we have over 30 drummers again.

Jake says to the players, 'All of you who are new, just copy someone that looks as if they know what they are doing. We're going to go on a walk. If you'd like to join us you'll need about five hours.' People think he's joking.

The plan is to play all over London with our 'Climate Emergency' banner. It seems the police don't mind the band as long as we're moving. You can't tell someone to 'move on' if they are already moving on.

We march to Victoria where three police appear. They are going to facilitate our playing. As we march around Hyde Park Corner they walk beside us to 'contain our procession' or something. Behind our 'Climate Emergency' message we march along right next to Buckingham Palace Gardens. When a road comes up on the left, the banner bearers swing around and block the road so that drivers wonder why they have to stop. We pass Victoria station and Channel 4 Television. The rope from the drum digs into my shoulder no matter how I try to adjust it. We reach Vauxhall Bridge Road and, as we march past residential properties, the windows fill with people watching us. Some blow kisses and wave. Some run out of cafés and shout 'Thank you'.

Where are the 'angry Londoners' that the media keeps writing about? Maybe they are all in cars. We turn down Vauxhall Bridge Road. I look up and there is a blonde woman standing on a high balcony. She has her hands clasped together in the 'namaste' or 'thank you' gesture. Her look of gratitude and love is so clear it's almost like an energetic rainfall showering down over us. I don't believe in all that 'energetic blessing' woo-woo thing. I'm clear that I don't believe in it, ha ha. But if there is such a thing – that's what she is giving us.

The police are marching along beside us looking cheerful. As we reach Vauxhall Bridge, a double line of police forms. They're obviously worried that we want to 'take' the bridge – to march into the middle of the bridge so that no traffic can get over it and then play for an hour while hundreds of people appear from nowhere and start to picnic. But no, we have no such ambitious plans. We turn left and play samba down Millbank as far as Lambeth Bridge. We're playing, 'We like potatoes. We like them mashed.' This is what the radical troublemakers are thinking of as we march along. Mashed potatoes. I haven't had time to eat but a mixture of drumming energy and adrenaline seems to be an adequate substitute for food.

We reach Lambeth Bridge and again the police scurry about nervously, but we don't have our sights on Lambeth Bridge either. We end our session with a dramatic crescendo then walk silently and respectfully down the little footpath next to the bridge for a rest in Victoria Gardens. It feels very good to take off the drum and sit on the grass. We drink water.

Three policemen wander over. 'What are you doing next?' I hear a policeman ask one of our more, er, 'interesting' members. 'We are very worried about flowers,' he says. 'We're planning a discussion about that.' The policeman smiles, but I didn't want this man to be cheeky or rude. At the moment the police seem not to mind the samba band and I want it to stay that way. I saunter over.

'Hello,' I say. Friendly as a Labrador.

'Who is your leader?' The policeman asks me.

'We don't have one. We have a musical director, but he's just a young lad who teaches drumming. Can I help you? I don't know anything about anything.'

'We're just wondering where you're going next.'

'That's interesting,' I smile, 'because I was wondering the same. I rather hoped we might stop at Vauxhall as I can get a bus home from there, but it seems not. This is a democracy so making decisions takes a long time. May I ask you some questions?' I smile again.

'Yes, you may, but I'm just letting you know that I'm putting my camera on. If that's OK?'

'Oh yes – I love cameras. I'm an author. Free publicity.'

I speak to the camera directly. 'My name is Isabel Losada. My books are available at all good bookshops and on Amazon. But personally I prefer it when readers buy them at independent bookshops as I'm a supporter of independent booksellers. My best-known book is *The Battersea Park Road to Enlightenment* which is also where I live – if you're interested. I write narrative non-fiction.' I turn back to them. 'Will that do as an introduction?' I think I'm supposed to worry about the camera.

'Nicely,' one says.

'So, I want to ask you. You appear to be both looking after us and arresting us?'

'Our job is to "facilitate your peaceful protest".'

'Whilst preventing crime and disorder,' says another.

'And to protect life,' says the third.

'Ah, yes,' I sigh. 'But only human life, right? Your job doesn't extend, sadly, to protecting animal life – or bird life and insect life. Look [I use the line that had been used on me at Hampstead Heath] no birds.'

We look around. There are no birds to be seen anywhere. Not even a pigeon.

'We like nature too,' one of them says.

'Course you do. I just wish your mission statement "to protect life" extended beyond the troublemaker species and then you'd have to change sides and play drums. Anyway, thank you for this brief interview. I must see whether the band has made any decisions now. If you want to be helpful then you could kindly wander away and stand by your van – that would facilitate our peaceful conversation about what we're doing next.'

Suddenly, there's a shout amongst the drummers. 'Bloody hell,' says Jake. Someone in Parliament Square has sent us a picture of police reinforcements arriving to 'clear' Parliament Square. There must have been 200 or more police in a line, two abreast, marching in. 'They're arresting anyone they can and dismantling the gazebos.'

'Hey, look at this!' Someone calls over the three members of the police we've been chatting to. We pass the camera to them.

'They seem to have sent in some reinforcements,' says a policewoman. 'Thank God for that. I've just done a 12-hour stint. They could have told us.' And they wander back to their van, presumably in the hope that they can now go home.

We gather nervously into a tight group. A message comes from Parliament Square. 'They've cleared one of our roadblocks. They're arresting everyone. Don't come in. You'll all be cannon fodder. We don't want the entire band arrested. We need you.'

'They say that they don't want us to go in. Those are our instructions,' says Jake.

'OK, so where now then? Waterloo Bridge?' someone asks.

'How about the Shell Centre? I'd really like to make some noise there. Polluting, planet-destroying bastards,' says someone else.

'Or maybe we could just go home and meet tomorrow?' says a young girl with a huge base drum. She's obviously exhausted. Me too.

'Hold on – we have another message,' says Jake. 'It says: please wait so we can see what happens in the square.'

Then immediately: 'They've changed their minds. They'd love us to come in right now but they say we may all be arrested so we should talk about it first. OK,' Jake says. 'All action is voluntary. No one has to come in if they don't want to. Everyone at Parliament Square has been issued with Section 14 notices so it's now an illegal gathering. Anyone can be arrested so there is no shame in leaving. As I've said before – we're not "arrestables". There are plenty of others who have thought about this in advance and are willing to be arrested. If you have any reason for not getting arrested and may suffer long-term consequences – please leave now.' He waits. He looks around.

My mind races. I had decided not to get arrested. I've heard that if you have a criminal record, the USA will never let you visit again. For all its faults, and the flight, I love America. And Japan doesn't let you in either. My precious taiko-drumming training – I'd never be able to meet the Japanese masters. But is it true or just scare tactics? And what are the other implications? I haven't thought this through.

'I really don't want to be arrested,' says a girl.

'Me neither,' says a young man. 'I just became a father.'

Jake says, 'None of us want to be arrested. If we did we'd be among the "arrestables". But we're asked to go in. If you want to leave, leave now. We are all free. No shame, no blame, remember? So – anyone coming in?'

Every single band member gets up and put their drums on. We're all scared and a bit sad, knowing it's suicide. That picture of rows of police is obviously designed to intimidate. And it works. But not well enough.

'Come on, Isabel,' I speak to myself firmly. 'Pull yourself together. They're not armed – this is Britain. You're not going to get shot. Just be thankful you live in a democracy.'

Jake blows his whistle. 'What do we want? Bunny ears!' And we march off in the direction of Parliament Square. When we get to the road block we play a few fast breaks that start soft and get louder. Zoë – who I'd met at the introduction and who was in the square – said to me later, 'There was so much happening. There were random arrests and everyone was panicking and then suddenly we heard this incredible noise: it was like rolling thunder and we thought, "What the fuck is that now?" Then it went Boom Boom Boom and we realized it was our samba band. It was fucking awesome.'

I wonder whether the police will let us into the square but they do. The people there cheer up and start dancing. We play as well as we can – and joyfully. The police look at us. Maybe their instructions to arrest everyone doesn't included band members. No one comes up and tries to take our drums away. No one tries to caution us. Mind you, it's quite hard to caution a drummer who is wearing ear-plugs, hitting a drum and focusing her eyes on directions from the musical leader. You can take away the leader but then another leader will spring up. They have learned this much. In the confusion, a roadblock which had been lost is retaken by fresh arrestables.

A man beckons us to come and play where people are locked down. So we march over and play there. Jake's energy is amazing. To conduct an inexperienced and motley samba band takes huge leadership, let alone under these conditions. He crosses his arms to change tunes and he waves them frantically to keep the counts correct on the breaks between one tune and another. He walks up to each section to correct what we're playing. In the middle of the chaos this poor man is a musician and wants us to sound good. It's so easy for a large group of drums just to sound like a cacophony – but Jake is determined to make us sound like musicians. Someone needs to give this man a medal.

We stop for five minutes, look around and the 200 or more police reinforcements have vanished. There are small groups carrying people away. Where did the rest go? Spirits of the activists are up again. They sing, 'Power to the people. The people have the power.' Skeletons of animals that the artists have made dance on top of poles. Beautiful white bone figures of animals that have become extinct dance with them. And human skeletons too – white bones dance against the evening sky.

We march round the square again and then, just when I think Jake surely must have to stop, he leads us all round again. Perspiration pours from his face and all of us are beyond exhausted. We have been playing on and off for seven hours at this point. Then finally he moves to lead us out of the square next to one of the roadblocks that has been left open for pedestrians and cyclists.

A police van moves to block the pavement. 'Oh fuck, we've kettled ourselves,' I think. That's it. We're done for. They're just waiting till we stop playing to serve us all with a Section 14 and then they can arrest every single one of us. The band exchange panicked looks. We go on playing. And then Jake realizes what's happening. The police van has mounted the pavement to move off the road – they are opening up the roadblock to let us march out. I thank all the gods for whichever duty officer is in charge that night. We march right out of Parliament Square. Behind us the activists are cheering.

THE BIGGER PICTURE 2
Are You an 'Arrestable'?

So, can the people get the government to listen? The jury is still out.

By Friday morning practice, our numbers have doubled. The atmosphere of a car-free Marble Arch is so relaxed. Seeing London car free makes you realize just how much the noise and fumes from cars ruin the quality of our lives. Apparently, the pollution levels around Marble Arch and Oxford Street, which are normally above legal limits, are down by 45 per cent and, breathing here, I can smell the difference.

A teenager walks by with a homemade banner: 'The greatest threat to the planet is the belief that someone else will save it'.

That's it. I thought. That's it. That's the threat.

I can't count all the tents. There are children's activities, talks and people discussing climate change and how to get the government to take action. This is an alternative world, one in which people of all ages, backgrounds and social classes are here to talk about living differently and taking care of our planet. And there is so much kindness and generosity here. On the stage people are singing and there is a feeling that together we may be able to turn this potential destruction around.

We play drums all morning and by the time we reach Oxford Circus, the police have formed two rings around the pink boat and are removing individuals who are stuck to the bottom with superglue or who have their arms locked in metal pipes. The police have to cut through the pipes with hand grinders that are in horrifying proximity to human flesh. It must be terrifying for both those using the grinders and those being released. Three of the last to be taken out are a 14-year-old girl, a man in a wheelchair and a woman who is 7 months pregnant. What courage.

Activists are singing. The police look relaxed. I suppose with so many people wearing labels and holding banners saying 'We're non-violent' and 'We respect the police', it must be an interesting gig for them. A group of students sing 'Power to the People' and we all sit and watch the mast come down from what is now called 'our beautiful boat'. It's named the Berta Cáceres after the Honduran environmental activist who was assassinated for her environmental work.

'Boo!' shouts the crowd. Do the police really believe in what they have to do? It must be hard if you are a professional policeman or policewoman who has read about climate change. As far as I can see the only argument for clearing the streets is that these actions must make it impossible for the disabled to go about their lives. On the other hand, there are a lot of Extinction Rebellion members with disabilities who point out that if the governments of the world don't make substantial policy changes, there will be no life for anyone.

An atmosphere of sadness descends upon the crowd. Instead of allowing us to move the boat to our legal camp in Marble Arch, this beautiful symbol of the moment is confiscated in order to be 'used as evidence'. Ha ha. As if pictures and film of the boat in the middle of Oxford Circus wouldn't be enough for them to prosecute – they want the actual boat. Then someone in the crowd shouts, 'We have more boats!' We sit in the sun watching the police move the boat a couple of hundred yards before being held up by further human roadblocks. My respect for this organization – or disorganization, or whatever it is – is growing by the second. Could I even have imagined, a couple of weeks ago, a campaign of civil disobedience of this size? As I sit in the sun, over 700 people have now been arrested and the number is rising by the minute.

A message comes through that the citizens' assembly has voted to meditate. For meditation they don't want the samba band so we play our way back down Oxford Street to the larger camp in Marble Arch. It's now become a huge playground. Some boys on bikes are showing off their wheelie skills. People hand out leaflets and explain why we're playing. One final show-stopping number to end and we're done.

I need a cup of tea and a chat. I call Zoë. 'Are you still here?'

'Yes, I'm in the welfare tent.'

'I'll come and find you.'

I find tea and a free vegan stall, and go to look for the welfare tent.

'So, how's it been here?'

'My job has been to give people tea and chat. I see you have tea.'

'Yes, but I'd love the chat. Can you chat with me?'

'Yes. Unless someone comes by in distress for some reason and then I'll abandon you. I also have to go round and ask people with beers to leave the site. I'm the nice police.'

'How's it going?'

'It's intense.'

'Yes.'

'But in the most beautiful, emotional, moving, powerful way.'

'What have you been doing?'

'With the wellbeing team I've mainly been handing out water, tea and biscuits to people on the front lines at the roadblocks and checking they're OK. I've been at lots of arrests.'

'I haven't been close to the arrests – being with the band. How's it been?'

'The Parliament Square arrests were the most moving for me. A group of Shakti singers gathered around to support the arrestees and the crowd joined in, singing, "This land is all that we have for sure" as people were carried off.'

'I feel sorry for the police. Honestly, I do.'

'I was standing next to a woman whose husband and daughter decided to get arrested. They were the first to be taken off and she had tears in her eyes and said how proud she was of both of them.'

'What a week.'

'I also saw a young teenage boy have his hand unglued from a lamppost. He can't have been much older than 14 or 15. To care that much about the environment, at such a young age, was so moving. I can't believe he was willing to give himself to such a big cause that young.'

'So impressive.'

'Last night I also helped a young black man try to locate his belongings after he'd lost them whilst being arrested. Considering he had lost his Deliveroo bag and ID card he seemed very calm. He'd seen the protests randomly on

Monday while he was working and he was so inspired that he decided to allow himself to be arrested on Tuesday. He kept talking about how much love and compassion we all shared for one another and how he felt truly listened to in the Extinction Rebellion spaces.'

'The listening has been amazing.'

'The other thing that has struck me again is just how lucky we are to live in this country. Even simply being able to take a week off from work, like I have, shows that we are privileged. And the colour of our skin means we are not discriminated against.'

'Sadly, it's true that it's easier to be a protestor if you're white. I can't believe we still live in a society that makes decisions about people based on the colour of their skin rather than their character.'

'The black protestor was the only person of colour I met who was willing to be arrested.'

'It is very white at Extinction Rebellion. We've had two or three people of colour among the drummers but it is 70 per cent white.'

'There were other black or mixed-race people I spoke to who said they would be too scared to protest in the way that we are because they are afraid of police brutality and the repercussions of having a criminal record.'

'I can understand that. It's completely unjust.'

'But we are most privileged for having the ability to have our voice heard in this way. There are so many other countries around the world where doing far less than we have would end in serious violence or potentially being shot. Yet here we are, with over 800 people arrested and no one with any physical scars.'

'Let's hope it stays that way in the years to come.'

'There's been enormous personal sacrifice. I've been blown away by the fact that so many people are so passionate about this cause that they've been willing to sit in roads, glue themselves to pavements or stick their arms in lock-ons.'

'Yes.' We sit in silence for a while. Then I ask, 'So you've had a good week then?'

'I'm exhausted, but it's been utterly inspiring.'

We stay at Marble Arch as the sun goes down and the moon comes up and listen to someone play Bach on a glockenspiel. Without the cars, we can hear every harmonious note.

The press coverage hasn't improved as the action continued. I wish I could say that it had.

Being involved in something like this lets you know just how much the media lie. And not only do they lie but they are just mean. Or at least the way these stories are reported lacks inspiration, humour, joy and any kind of positive interpretation. They don't give the activists the benefit of the doubt. Only in the *Guardian* could I find balanced stories and one with a list of interviews with protestors asking them what they are doing and why. Most of the media I saw were so negative that they didn't reflect the actual events in any way.

One pile of vitriol by Leo McKinstry in the *Telegraph* is headed, 'The Brain-Dead Eco Snobs Causing Mayhem on Our Streets are Beyond Parody.'

Someone called Ross Clark writes in the *Spectator*, 'Extinction Rebellion is a Wannabe Marxist Revolution in Disguise.'

Sky News gives an interview with Cressida Dick, the Metropolitan Police Commissioner, in which she says, 'People are losing money.'

The London papers even go into great detail about how the police move the 'troublemakers'. The *London Evening Standard* actually uses the phrase 'a small group of protestors'.

The Times writes, 'It took more than 200 police officers, a team of engineers, logistics experts and a mountain of heavy equipment to remove a small pink boat from Oxford Circus yesterday after a stand-off between the authorities and climate change protesters that lasted five days. The boat had become a symbol of the mass protests led by Extinction Rebellion that have brought the capital's roads to a standstill.'

The newspapers talk about the 'inconvenience' caused to shoppers. They talk about the fact that extra police have had to be called in from around the country. They talk about the fact that the demonstrators are mainly middle class and polite. (Politeness is now apparently a bad thing.) They talk about the fact that police cells are full. They say that somewhere (heaven knows where they found this or maybe they made it up because it certainly wasn't my experience) a passing motorist had become so angry that he had punched a protestor.

But they don't talk about the amazingly positive relationship that has been created between the protestors and the police. They don't make clear the inspiring commitment of the young people and their passion to create a life for themselves and for future generations. They fail to mention that the campsite has remained 100 per cent alcohol and drug free. They don't seem able to imagine that something remarkable is happening. That never before in the history of the environmental movement have so many people stepped forward to give such a clear signal that policies need to change. They don't run articles on the unprecedented drop in air pollution in London due to the lack of cars. None of the papers, as far as I'm aware, thinks this is worthy of a mention. They don't talk about the amazing bravery and courage of grandparents who have glued themselves together or onto pipes so it's hard to remove them from roadblocks. They don't describe the profound levels of support from people singing in groups around those who have locked themselves down. They don't interview any of the now over 900 people of diverse ages and backgrounds about why they are prepared to get a criminal record. They don't ask why it is that, as police remove those locked on, hundreds of others wait to take their places, singing, 'Love, Peace and Respect'.

And, above all, of course, they don't write about climate change.

I feel rage on behalf of our beautiful planet and every single arrestable soul who has shown such courage. It's hard not to feel discouraged. Bastards.

The following morning I get up early, drink some tea, pick up my drum, and, like all of us that are out there, I go back. I put on my Extinction Rebellion badge because I'm proud to be amongst these people. These crowds that contain so many individuals that are so much braver than I am.

When I arrive, there are already newcomers learning the rhythms that I now know. As we practise, two new musical leaders say to the crowd, 'If you'd like to join us – pick up a drum and listen.' And new people step forward. I guess to anyone in the crowd with any experience of drumming, the magnetic attraction of the rhythm is irresistible. More people step forward and pick up drums. We are now a group of 52.

Then the long process of democracy begins. 'Dialogue and not debate,' we are reminded. One of the drummers says,

'We are here to model the citizens' assembly we are asking for. We ask questions and when we are ready, we vote.'

We sit in a circle to hold a citizens' assembly. People hold up their hands with various signals: a C for a 'clarification', a P for a new 'proposal'. If we like what someone says we do jazz hands in the air. If we don't like it we do jazz hands low down. Then there is a circular 'could you wind up?' signal. It's easy to see why: in an empathetic democratic system that wants to value everyone's viewpoints equally, it takes a very long time to decide anything.

'We want to learn how to make decisions better without shouting others down.'

Lots of jazz hands in the air.

What we are exploring, although slow, is certainly a better model than the one we see when we watch the braying tribal slanging matches that pass for the democratic process in Parliament. We are also asked to be mindful of when to speak and when not to speak. Those who haven't spoken before are given priority. Sitting in a circle there is no sense of 'us' and 'them'. No one has to shout 'order, order' because there is no derision. There is a genuine chance to listen and to think before speaking. Everyone is heard.

We eventually vote, 'walk silently through the park and then march to Parliament Square', but this time without annoying the police or causing roadblocks. As we walk through Hyde Park we're sad to see the park covered with litter and plastic bottles. Of course, it isn't the climate activists who have left single-use plastic. Unfortunately for us it's the result of an annual event in the park where everyone comes to smoke cannabis en masse, which has taken place at the same time. We stop our walk through the park several times to pick up as much litter as we can in the certain knowledge that someone will post pictures of this and blame the climate activists. (And later – sure enough – that's exactly what happened.) I like litter picks. Instant results. The police watch us. We don't appear to be troublemakers.

All day we march and play. At dusk, when we arrive back at a packed-out Marble Arch, a treat is in store. Greta Thunberg has arrived. We move forward to sit on the ground as a diminutive but powerful teenage girl comes to the microphone. Her address is brief and clear. She thanks everyone for the privilege of speaking to us and then says, 'For way too long the politicians have got away with not doing anything about the climate crisis and the ecological crisis, but we will make sure that they will not get away with it any longer.'

Thousands of people cheer.

'Humanity is now standing at a crossroads. We must now decide which path we want to take. What do we want the future living conditions for all species to be like? We have gathered today and in other places around the world because we have chosen which path we want to take. Now we are waiting for the others to follow our example.'

Another huge round of applause.

'And we will never stop fighting. We will never stop fighting for this planet and for the futures of our children and grandchildren.'

Impressive? She is. A wall of press is there. There is a feeling that history is being made.

The following day I go to hear Greta Thunberg again, speaking at Friends House with Anna Taylor – her UK counterpart – and the MP Caroline Lucas.[11] Guardian Live livestreams the event. Something has shifted. The event is sponsored by the

UK Student Climate Network, Greenpeace UK, Amnesty UK, Campaign Against Climate Change, Friends of the Earth, 350, National Union of Students (NUS), Policy Connect, 10:10 Climate Action, Parents4Future, Quakers in Britain, the All-Party Parliamentary Group (APPG) on Climate Change and the World Wide Fund for Nature (WWF). In other words: a very impressive line-up.

It's good to sit back and listen to them. Caroline Lucas is such a warm and generous woman. She shines.

'I'd love to move to Brighton Pavilion just so that I can vote for Caroline,' I say to the man in the seat next to me.'

'I think we've all considered that.'

Anna Taylor is so clear. They are all so clear.

At one point in the Q&A someone asks Greta, 'How do you deal with a climate change denier?'

She says, 'I don't.'

Afterwards, I bump into Alex Thomson, the Chief Correspondent for Channel 4 News. He says, 'My children are always asking me, "Why don't you do more on the news about climate change?" It's all they care about. My boss says the same. His children don't care about politics, they care about the planet.'

'Good.' I smile. 'Please send my love to your children and ask them to keep on asking this question.'

'I will.'

'So what are your thoughts about this week? About this action? All these arrests? Is Channel 4 concerned about inconvenience to Londoners?'

'It's an unprecedented and extraordinary victory ... 1,065 arrests? And that's the figure released by the police. Do you realize that the last time there was a non-violent uprising – civil disobedience of this kind – was with the Chartists in 1839?'

'So you think, from your point of view as a full-time news reporter, that this has been a success?'

'Isabel, it's "an historic success". No one was talking about climate change and ecological breakdown. Now everyone is. This has completely changed the conversation.'

Finally. This feels good. This feels like a step in the right direction.

BIRDS 1
Vegan Fat Cakes and the Merits of Pyracantha

Traditional gardening, conservation and birds: entire subject areas that fill me with rage.

What most grieves me is that the UK has an appalling record in biodiversity loss. We are world leaders in killing off species. We are one of the most nature-depleted countries in the world.

I suppose we have to accept that there are many aspects of this that are beyond our control. It's that 'Serenity Prayer' again. The first part: 'Having the serenity to accept the things I cannot change' has never been my strong point. Nor the third: 'Having the wisdom to know the difference between what we can change and what we can't.' Wisdom? Even the word is kind of dated, let alone the quality.

But 'courage to change the things we can'? At least we can have a go at this, can't we? I mean – even if we fail we can give it our best shot.

So, this decline in the bird population – it's a disaster. We can all do something about that. As we now know, we need to look after the insects first. And we need to look after our soil so that the insects thrive. But on top of that we can take some actions specifically to help the birds. We can all do something. Even if you don't have a garden you probably know someone who has.

Advice on the website of the Royal Society for the Protection of Birds throws up many ethical dilemmas for a vegan like me.

If you look up 'feeding the birds', you learn that 'suet' balls are a favourite bird food. Do you know what suet is? It's the raw hard fat found around the loins and kidneys of cows or sheep. Ah, the irony of there being yet more exploitation of animals in order to feed birds.

The RSPB – and many pet shops and garden centres too – sell:

Coconut shells packed with premium suet. (I can understand why they don't call it 'cow and sheep fat' – it's not such an easy sell.)

Super suet bars in three different varieties. (Honestly.)

Hanging suet cakes.

Nope – not with my money. I want to look after birds but I'm not buying suet. No flipping way.

I search the internet for alternatives and find a site called 'Birds don't eat cows' offering a great range of non-animal-based bird foods – but it's in the USA. I ring my local pet shop. Same old, same old. I can't find anywhere on my continent offering fat balls not made of suet. I ask the RSPB and they tell me, 'We're looking into this. You're not the first person who has raised this.' Good. Watch this space.

On to the next dilemma … It's best not to be a Buddhist who has taken a vow not to harm any living things. I mix in some strange circles, but I know a surprising number of people who have taken this vow, and many more who haven't taken a vow but who have a deep desire not to hurt any sentient creatures. So, it's not good to discover that the next most popular bird food is mealworms. You can buy them dead or alive. Mealworms are beetle larvae that are farmed and killed in vast numbers before they have the chance to turn into beetles. I ask around a bit and a friend says, 'I know a rep for a bird-food supplier who supplies garden centres and pet shops. Their supply of mealworms comes in from China in those huge transportation tankers and then they are packed here for shops. They are all farmed. It's grim.' Bird lovers sprinkle both dead mealworms and live meal worms on bird tables and on the ground. I wonder what the ecological impact of putting live mealworms on the grass is. There surely has to be one. So I'm not going to be buying mealworms – dead or alive.

I buy some 'favourite blend bird food mix' from the RSPB. 'Only the very best for your wild birds' it says on the packet. From the picture I think it's seeds, but I don't read the small print and when it arrives I find it's made up of 'sunflower seeds, suet and dried mealworms.' So much for my efforts to be conscientious.

'Maybe I can make something and feed the birds myself?' I think. I find a video online about how to make vegan fat balls. This looks like a good sticky activity if you have children of the right age who need diverting from screens. Make your ball either to drop into a metal bird feeder or around a piece of string to hang. I have to try this once – just so that I can let you know whether it works or not.

Here you go: how to make a fat ball for your birds without feeding them cows or sheep.

Put these five ingredients into a bowl:

50g of some kind of vegetable-based shortening – (I found Stork easily in my local supermarket)
2 tablespoons of peanut butter
70g polenta (found this easily too)
2 tablespoons of wholemeal flour
1 bag of wild birdseed

Mush everything together. At first you think it's not going to work and you're just going to have a sticky mess, but then amazingly you find that you can shape it into balls. Ta-da!

I did this today and the entire process took 10 minutes and 24 seconds, including cleaning the bowl, washing my hands and sending a photo proudly to my daughter who complained that as I don't bother to cook for myself efficiently it seems absurd that I now have four birdseed balls hardening nicely in the fridge. But I'm very proud of my non-cow-or-sheep-based fat balls. They will stay in the fridge overnight and tomorrow be presented to the garden birds. Next time I have a coconut I'll make some more, stuff half a coconut shell with this goo and then hang the masterpiece out on a string for birds to play with while hanging around feeding themselves.

I appreciate that you're busy running your life and many of you will be deciding at this point that I have way too much time on my hands. So, if you don't have children and don't want to get greasy hands, here are some easier ways to feed the birds – compliments of the RSPB.

In spring and summer you can feed them soaked sultanas, raisins and currants. You can put out soft apples and pears. (Never throw bruised fruit in the compost – cut it up and give it to the birds) You can also put out bananas and grapes cut in half. Do not put out bread. Peanuts are not a good idea at this time unless you put them in a bird feeder where they have to peck them out in little pieces. Whole peanuts fed to fledglings could kill them.

In the breeding season, blackbirds and song thrushes are eating worms while tits and chaffinches are eating caterpillars. If it gets very dry the birds can't feed from the ground so just keep an eye out and be aware of your birds. In a very hot spell in summer, sometimes the most useful thing you can do is water the grass for a while so that the worms come up and the birds can feed.

In the autumn and winter, as well as all the above, birds really need fat to survive the cold so that's the best time to make and hang some fat balls. If it's really freezing, put out food daily. Also providing fresh clean water all year round, both in birdbaths and at feeders, is as important as food. Clean your bird-feeding stations regularly or diseases can spread – and make sure all this is kept well away from any feline friends.

Are you thinking that all this seems a little insane? Me too. While I acknowledge that feeding birds has been a huge part of keeping garden populations thriving, the more I think about it the more the whole venture seems out of balance. I have included the above in case you want to have a bird table, but birds shouldn't have to depend on humans remembering to put out bird food. Nature knows how to feed birds.

What we need to do with our gardens is make sure that we choose shrubs and trees that feed birds. We need to plant fruit trees so that both humans and birds have food as well as trees with berries.

Looking at the RSPB website, I couldn't quite work out why it didn't have a list of berry-bearing trees and shrubs across the home page. Surely they could sell these (marked up to help their profits) on the bird food page alongside the suet balls. Perhaps I'll suggest that to them too. They do say that they welcome feedback. I wonder just how much feedback they welcome.

Meanwhile, I've done a little of my own research – so those of you that have a garden or a friend or neighbour with a garden or feel like doing a little guerrilla gardening, this is a better way to feed the birds. This is what you want in your garden to feed the birds:

1. If you plant barberry, which is a deciduous shrub with amazing bright-red berries, then in winter you will have food for thrushes, fieldfares and redwings.
2. If you put contoneaster in your garden somewhere, the flowers attract bees. It is used as a larval food plant for five different types of moth (and moths, of course, feed bats) and the bright-red winter berries are food for thrushes and waxwings. You could even make a hedge from this wonderful plant. I mean, why put up a fence when you could have a contoneaster hedge?
3. Now all you literary types, I really feel you ought to have common hawthorn in your garden – as hawthorn is one of the favourites of Proust. For Marcel, the sight of a hawthorn in flower in May was so unbearably beautiful that he was sometimes forced to look away. Hawthorn also has an amazing perfume which Proust describes as having the 'bitter-sweet fragrance of almonds'. Hawthorn grows slowly in glorious white bushes and can also grow up into trees with wonderful gnarled trunks. As well as having all these wonders, the red berries in winter provide food for starlings, finches, crows, blue tits, thrushes and waxwings. And it's cheap

too. I bought four small hawthorn plants yesterday for £6.95. I hope you're as excited as I am. I mean really – joyfully looking after the planet by planting hawthorn may be something small, but if you're a hungry thrush in the snow it could be life or death.

4. There are many different kinds of ivy. One that is called *Hedera helix* or sometimes European ivy (if you're that way inclined politically) or English ivy (if you're more politically inclined that way) or – if you want to avoid politics altogether – you can just call it ivy. *Hedera helix* is the posh name for the common variety that many gardeners pull down because it can be difficult to eradicate and it crowds out other plants where it is established. However, it has black berries in autumn and winter that are food for wood pigeons, collared doves, waxwings, thrushes, jays, starlings and finches.

5. Common holly. Mistle thrushes love the berries and you can save money at Christmas by just bringing a branch or two into the house. I had a holly tree at my old house and even though the leaves were prickly, I still miss it. Strange how you can miss a tree.

6. Honeysuckle (or *Lonicera* if you want to be posh and use the Latin name) seems to have all sorts of benefits for a garden. There are many different types and they are all glorious with highly perfumed flowers. They are easy to grow, pretty much indestructible, and not prone to pests or diseases. If anything, the only problem with them is that you have to keep an eye on them and cut them back occasionally. Otherwise, if you turn your back they will have doubled in size. Of the many different types, *Lonicera periclymenum* has red autumn berries that are food for (how's this for a list?) robins,

blackbirds, song thrushes, garden warblers, tits, crows, finches and waxwings.
7. If you live in an area where you have security concerns to the point where you or someone else has had to put up barbed wire or broken glass to keep out intruders, you could consider getting rid of all that and planting pyracantha. This plant has thorns that are so lethal that the old gardeners where I currently live (my garden is communal) refused to cut it back or go anywhere near it, as they said that they were not insured. If someone were to fall on this plant it would do just as much or more damage than it would if someone fell onto barbed wire. In many ways it's not what you'd call a 'nice' plant. But it produces abundant vivid-orange fruit in autumn and winter and the wood pigeons and thrushes just love it. I often watch the wood pigeons eating the berries and wonder how it is that they don't spear themselves on the thorns. But they never do. This plant should be so well known that any self-respecting burglar would take one look at it and say, 'Forget it – they have pyracantha.'
8. Red-berried elder, also known as red elderberry, is good if your ground is very wet because it thrives in those conditions. The stems, roots and leaves are poisonous for humans but butterflies love the flowers while waxwings and thrushes eat the autumn fruits.
9. Whitebeam is native to southern England, so if you live in southern England this is a good choice as we're all supposed to be planting native plants. According to the Woodland Trust, it's also widely planted in the north of England. In the north-west they call the berries 'chess apples' and humans can eat them when they are nearly rotten. The flowers are food for pollinators, the

leaves homes to at least four species of moth, and the scarlet berries, which ripen in late summer and early autumn, are food for wood pigeons, fieldfares, redwings, blackbirds and mistle thrushes.

10. And finally – if you live at high altitude or up a mountain, there's the beloved rowan, which I planted so much of with Trees for Life. It's native in the Highlands of Scotland but is so much loved that it's also planted just because it's beautiful. The leaves are eaten by caterpillars of moths and the caterpillars of the apple fruit moth feed on the berries. The blossoms produce food for the pollinators and the berries feed blackbirds, mistle thrushes, redstarts, redwings, song thrushes, fieldfare and waxwing. Rowan is also good for keeping out witches and evil spirits, which, you never know, is a quality that you may welcome.

So, I hope this has been inspiring or useful. I know that not everyone has a garden or any outdoor space but hopefully everyone knows someone who has, and who may welcome ideas and inspiration. Now I'm obviously not saying that there is anything wrong with having a bird table in your garden or on your window ledge – of course not. I just wanted to offer this little list as a more long-term option. And, of course, planting trees also helps the atmosphere and gardening, in any form at all, and has been shown to be good for mental health in humans.

Maybe there is just one reader of this book who has a very large garden and who will decide to plant all of these. It will create a natural feeding centre for birds at the very time of the year when their food is the most scarce. And somehow the birds will find it. They will come.

MONEY 1
Putting Your Money Where Your Values Are

Talk about not joining the dots. I've just never really thought about it. I've been banking with NatWest since I was a teenager, but I suppose I've always known, somewhere in my mind, that the highstreet banks do all kinds of things that I wouldn't necessarily agree with. Some good, some . . . er . . . not so good. Yet, I've always accepted it as one of those unavoidable necessities of the world we live in. I have to have a bank account so I may as well have one that I can trust and I know works. But what has my trust been based on? Purely the fact that the banking system seems to work. They have never lost any of my money and as I've never had a large amount of money I've assumed that it doesn't really matter who I bank with. There must be billions of people in the world with this kind of half-formed thinking. Imagining that the tiny amounts of money they hold in hightstreet banks (if we are fortunate enough to have accounts that are in the black) don't matter. It's that thought again, isn't it? My bit doesn't count. My vote doesn't count. My individual actions don't make any difference.

And yet a proportion of the money that we have 'in the bank' is lent to . . . well, do we have any idea who it's lent to? We don't. We've never asked. Or at least that is the position of many of us who bank with a high-street bank.

I wrote to my bank to ask them what they were doing with my money and they told me: *'We don't publish a list of all companies that we lend money to, however we do publish our environmental, social and ethical policies which clarify prohibited and restricted areas of lending across a number of sectors.'*

'Restricted?' Personally I don't want *any* of my money used to help petrochemical industries, arms industries or any organization that may be supporting those industries. I don't want to have to find out what 'restricted' means. I don't want to wonder why they don't publish a list of the companies that they lend money to. I don't want to have to try to understand how 'ring fencing' works. And what's more, I don't need to. I can vote with my feet and choose a bank that has already thought about this for me.

There are two banks in the UK that believe in exclusively ethical investment (as far as I'm currently aware). There is the Co-operative Bank which only invests ethically, and then Triodos Bank which puts a list of the companies in which they invest into the public domain. That looks like transparency to me.

Now, what has any of this got to do with the planet and becoming an environmentalist? Well – everything connects to everything. The petrochemical companies are very rich and make eye-watering – no, more – eye-stinging profits. Trillions of dollars brings them power, influence and more money. To keep their shareholders happy (millionaires need to be reassured that their investments will keep on growing), they even have to have reserves of oil so that they can prove they are going to be able to continue to make trillions of dollars of profit in the future.

So, while they go on digging for new oil, fracking some rock near you, taking the top off another mountain and generally putting more CO_2 and other greenhouse gases into the atmosphere than the planet can tolerate – who is funding them? Well, among others, probably me and statistically you. You are one of the solutions to putting an end to this behaviour. You can tell them that you are leaving unless they commit to only investing in renewable energy.

'But Isabel, I've only got £59.24 in the bank,' you reply. 'I hardly think that me leaving them is going to make an impact on the most wealthy industries in the world.' Well, maybe not – but that's no reason not to take your small action. In fact, disinvesting your tiny fund may be one of the very best things you can do for the planet and I'm going to tell you joyfully all the reasons why.

The Disinvest Moment is already powerful in the US and across Europe. Students have taken their universities to task for taking students' money and investing it in companies whose actions are killing off the very same future that the students are investing in. This has been so successful (university dons can understand the logic of 'we'd like a future please') that many universities have sold their shares in petrochemical companies. Church congregations have done the same. Community groups, cities, pension funds – basically anyone who has money has investments. By asking people to disinvest and to leave the institutions that are the planet destroyers, we remove their moral right to exist. This is what happened with the tobacco companies. They became known as purely bad. This now needs to happen with dirty energy. We only want to fund renewable energy. No excuses. No exceptions.

And once you start to ask questions and discover that the use of your money by high-street banks is not in the public domain, you can start to wonder what else you may have been funding. Perhaps you are a nurse unknowingly lending money to the arms industry. Perhaps you are a vegan unknowingly funding an intensive factory farm or that animal experimentation facility where they are breeding puppies for medical experimentation? Perhaps you are a teacher in a failing school, deeply committed to equality in education – and, unbeknown to you, you're helping to build the new wing of the super-expensive private school down the road? Perhaps you don't believe in private medicine but you've never looked too carefully at the names of the organizations your bank is lending money to. Do I need to go on? I don't think so. But I hope – like me – you are feeling both a little outraged and a little more powerful than you were before you started reading this. Enough, friends. We have the power to make a change here.

'But aren't all the banks just as bad as each other?' you ask. No, actually, I'm happy to report, with 100 per cent confidence, that they are not.

I'm with Triodos now. After 30 years with my previous bank it was so easy to move it was almost silly. I have no idea why I didn't move years ago. Triodos supply a MasterCard and, when I want cash, it comes out of all the same cashpoint holes in the wall that say 'free cash'. But they don't mean it. Disappointingly, when you take money out of the hole in the wall, they do take the exact same amount out of your account.

So, if you're up for a bit of desktop activism – why not email your bank and ask them who they lend money to, and if they don't give you an answer that's clear of obfuscation and ambiguity, leave them and join a bank with an investment portfolio that is 100 per cent in the public domain. Or in summary form: put your money where your values are.

EARTH 3

Befriending Your Urban Trees

It's quite an adventure learning about street trees in our towns and cities. Apparently, in the UK, 56 million of us live in urban areas while only 11 million live in the countryside. So this bit is for all of us who live in towns and cities. You can skip this section if you are one of those who has wisely chosen to live in the countryside. When a tree seed germinates it contains the potential for the tree that it can become. If you live in the countryside trees mainly grow freely to reach their potential; all you have to do is admire their beauty.

For the other 56 million of us, I thought I'd give you some brief thoughts on how to stay sane where the subject of street trees is concerned. If you are a dendrophile (such a great word) there are many forces to drive us to premature madness: the sight of heritage trees, which we should value more than we value our cathedrals, being felled for some questionable building project; the pollarding in the spring of trees that contain nesting birds (this is illegal but still happens); the cutting back of trees so severely that only what a friend calls 'angry fists' remain; trees that vanish overnight with no explanation given to local residents; saplings planted in planters that look good on architectural drawings but stand no chance of survival on the streets; young trees being suffocated by the too-tight straps attached for their support; empty watering bags everywhere – and of course, dead young trees. Well, it's enough to make you envy your countryside-living friends.

The government (whichever party is in power) loves to boast about how many trees it has planted/is planting/will plant. The numbers always look impressive – this local council has planted thousands of trees, this borough has an equally impressive number. It can feel like some crazy competition. My own borough, Wandsworth, in London, likes to call itself 'London's Greenest Borough', but at the time of writing it is still one of the London boroughs that hasn't submitted any figures to the Mayor of London, whose office is compiling the numbers. How we can claim to be the greenest borough when no figures are available is unclear to me.

You wouldn't imagine that I'd need to explain what trees do for us but it seems that some humans have had other things on their minds – so, briefly, this is why we need as many trees as possible. They clean our air and provide the very oxygen we need to breath, which is in increasingly short supply. Trees are beautiful and this beauty calms our nerves and contributes positively to our mental health. The roots of trees absorb rain water and so help us avoid flooding. Being able to see trees from a hospital window has been shown by researchers to speed recovery and it has even been shown that tree cover reduces crime. Trees increase the value of our homes; a street that is lined with trees is valued more highly than one without. As our hot days increase, our need for shade increases and the charity Trees for Cities has shown that the ground under the shade of a tree can be 8 degrees lower than under direct sunshine. Sometimes we don't even notice how much a difference trees make in our lives until we don't have them. I recently asked a friend who has moved to southern Egypt what she most misses about the UK. She replied without hesitation: 'The trees'.

But trees are hated by many local councils because, guilty or innocent, they can be blamed by insurance companies for subsidence in houses and if it can be shown that a tree is even implicated in a claim, it can be the council that has to pick up the bill. It doesn't matter if the tree isn't the actual cause of the subsidence – neither the insurance company nor the council want to spent time or money finding out. Easier to just take down a tree and see if that solves the problem. Trees are also over-pollarded because if they have too many leaves they need more water and this may lead to a claim. Better to over-pollard them and not take any risks. And of course the 'tree surgeons' (both good and bad) like as much business as possible all year round. I wish I was making this up.

So that's subsidence. Then of course mature trees are under threat from building contracts. Many councils are struggling financially. One of their sources of revenue is building – and what may stand in the way of building a block of flats? A 1,000-year-old yew tree with a protection order on it? 'Never mind,' thinks the building company, 'we'll just chop it down then say, "Oops – sorry – don't worry – we'll pay the fine and plant a sapling."' Sadly, this happens often. Sheffield City Council and Plymouth City Council are famous in the UK for the number of mature trees they have removed.

Then there is the danger of a tree randomly falling on you and injuring you as you go about your business. Some local councils seem to be particularly afraid of this one. So here are a few useful statistics from the National Tree Safety Group. I just love these. In an average year the number of people turning up in A&E departments suffering from injuries due to their encounters with the following objects is:

People injured due to footballs – 262,000
People injured due to wheelie bins – 2,200
People injured due to trees – 55.

You are 40 times more likely to be injured by a wheelie bin than a tree. This is a statistical fact that we can all enjoy.
And let's look at the possibility of death by tree.
Your risk of death by:

Injury and poisoning – 1 in 3,000
Traffic – 1 in 7,000
Trees – 1 in 10,000,000
Lightning – 1 in 18,700,000.

At the very least, we can say that using 'public safety' as a reason for cutting down and removing trees is a little exaggerated. If local councils really cared about public safety they could be far more effective by banning wheelie bins.

Our trees are not a danger. Our trees give us life and a better quality of life.

So how can a joyful environmentalist best help our trees? I have a range of plans. You remember Jenny's suggestion to 'be a little bit activist'? Here is a selection of activities. You can choose what you can most see yourself doing. The first is a wonderful and easy opportunity to take action that will upset no one. What you do is this. You simply decide that you will keep a vigilant eye on the trees in the streets around where you live.

Most councils are able to plant new trees but lack the resources to water them until they are established – which takes about three to four years. We can help newly planted trees very easily.

First, find the young trees close to where you live. Do they have a black plastic watering pipe placed in the ground next to them? This pipe means that you can water the trees roots directly. Whenever you leave the house and know that you will be walking past a sapling, take a large bottle of water and pour it down the watering pipe – or at the base of the tree if there is no pipe.

Or do they have one of those dark-green plastic watering bags? A word about those bags. This is just my opinion and the opinion of some other local dendrophiles that I know. They don't work. I played a game with myself for years to see if I could ever find one with water in. I played this game all over London in numerous boroughs and also in other towns and cities in the UK that I visited. I must have examined hundreds of bags and only once did I find one with water in.

Watering bags cause more harm than good because they give the impression that someone is putting water in them. They aren't. Instead people use them as litter bins filling the area immediately around the new young tree with litter so that when it rains no rain can reach the roots of the tree. Then even bags filled to the top with water only take about 8 hours to drain and it can be up to a month until someone charged with this job can get to that tree again. In short they are a disaster. Someone somewhere has made a lot of money from selling these green bags of plastic to every town and city in the UK. I have been (unofficially) advised by nameless authorities that the trees would be better off if we remove them completely from the saplings we choose to care for and quietly dispose of them. Find the watering pipe if there is one – unblock it and pour water down it. This is what a tree needs rather than a plastic bag over its tiny allocation of earth. Of course if you have a tree with a watering bag in your street and you are filling the bag yourself on a regular basis, that's different – you are using the bag as the designers intended. But anyone who is expecting someone else to fill it may well be disappointed and end up looking at a dead tree.

Some trees have a porous material that is placed right up against the trunk of the tree. If there is strange-looking pebbledash material planted up against the trunk – before you call your council and complain – try pouring water on it. Much to my amazement, trees planted by Transport for London on the Battersea Park Road red route where I live, have this around them. It didn't look porous till I tried pouring water on it. Then I smiled because Transport for London had taken better care of the new trees that I suspected. The surface protects the tree while enabling enough rain water to get through. Or that's the plan. They would still benefit with passers-by watering them. On the other hand – some very unenlightened councils (still looking at you, Plymouth City Council) have been known to put tarmac right up to the trunk of a tree so the tree stands zero chance of ever getting any water to its roots at all. I wouldn't have believed that humans could be that stupid but I've seen pictures of dying trees trying to grow in tarmac. All other planting methods, though, enable you to help your trees to survive.

The simple good turn of watering young trees is currently one of the most important and doable actions we can take in response to the old advice to 'think globally and act locally'.

And in case you're thinking, 'But surely you're wrong – the local councils do water them for the first three or four years until they are established?' – in theory they do, but it's complex. Councils lack resources, the job is often allocated to one exhausted man with an area impossible for one person to cover and, well, if you want your local trees to thrive, you may like to make sure they do by taking care of them yourself.

'But if the council is watering, isn't it possible that I'd overwater if I also water regularly myself?' you might ask.

No – a tree is not a houseplant standing in a waterlogged pot. If a tree is planted in the ground it's impossible to overwater it. So when you walk out to post a letter or pick up some oat milk, take a large water bottle with you.

On to the next level of tree support – should you wish to do more. If you find a tree that is dying or dead, needs a overtight strap adjusting, need re-staking or in any way needs more help than you can provide, you need to learn exactly who to go to and how to get a local 'tree officer' to appear and solve the problem for you. If you want to get good at this (like many things, it requires persistence and infinite good will) you may like to become an official 'tree warden'.

There is an organization in the UK called the Tree Council. Put the name into Ecosia and you'll find them. Put in your postcode and they will tell you if there is a group of volunteer tree wardens where you live. I know – it's the most ridiculous name – I've given them my unsolicited opinion about that.

The Tree Council and tree wardens are not political. They work closely with the local council regardless of who is in charge. Matters to do with trees have to be dealt with in close communication with councils because it's complicated. If you decide you want a tree outside your house, you can't just pull up the pavement and plant one because you don't own the pavement and there may be electric cables, water pipes and other reasons why a tree can't be planted there. So you need a tree officer who can find out these things and will also advise (we hope) on planting the right tree in the right place.

Sadly you can't request an apple tree on the pavement outside your house, because if a pensioner slips on a decaying apple they could sue the council. So you need the right tree, and one that has been shown to be able to cope with the levels of pollution on our streets.

This is why joining your local group of tree wardens is a good idea because the local tree officers (qualified tree professionals employed by the council), who are always overworked, will be more likely to listen to you if you have the network of tree wardens behind you.

The tree warden network across the UK has various groups of volunteers who do the kind of simple voluntary work that I've described above.

If you find that you have a local group, it's reassuring to know that you are not the only person for miles around looking out for the trees. If you don't have one then get in touch with the Tree Council and they'll help you create one. It's not hard. If you are fortunate enough to live in Birmingham, the group there became so large and so successful that they went independent. They offer courses to learn about trees and are a totally inspiring group. Put 'Birmingham Tree People' into Ecosia and you can see what's possible.

Of course there is no limit to what you can learn about trees: dendrology could engross you for life. But I'm assuming you are very busy, have a full-time job, maybe a partner who is often annoying, children or elderly relatives who would like more of your attention, maybe a hairy dog?

'Well, yes, Isabel,' you may well be thinking, 'and you seem to be suggesting that I take on the care of my local trees too?'

Yes, I'm suggesting this. But find manageable ways to enjoy this bit of action. If you have kids, learn what type of tree you are watering, maybe which year it was planted, why they chose that tree for that location. If you have a dog I'm sure he or she will be enthusiastic about a visit to a local tree. If you travel regularly to visit a relative, maybe there is a young tree on the route that you're already walking. If, like me, you need to get away from your computer more, here is a reason to go for a walk, complete a small action that makes a difference and walk home again. And I can tell you that this small action is infinitely rewarding. There are three trees local to me that I know would not have survived had I not been a little active on their behalf. To see a healthy established tree in a location where you know it will be likely to thrive, where once there was a struggling sapling, may not be much, but it makes me smile when I go to post a letter.

Looking after mature trees is a far move complex matter. Street trees, the trees in parks, or on public land all belong to the council rather than to the people. It's the council who can and sometimes will protect them. It is also the council who can cut down hundreds or even thousands of trees if it chooses to. There are various ways you can act to protect them. The first way, as I said, is to become a tree warden as that group, which is often made up of knowledgeable dendrophiles, may be able to help you get a Tree Preservation Order (TPO) placed on a tree. This, in theory, makes it less likely to be cut down.

There are all kinds of conditions. A council is unlikely to want to put a TPO on one of its own trees. So street trees are out – sadly. But trees on private land they will consider. So if you have a neighbour who is about to cut down an extraordinary tree that is in their front garden, and you and everyone in your street sees it every day and loves it, your local council can help you to protect it from your neighbour's saw. You are most likely to be successful if there is a beautiful tree in the front garden of a private house or a block of private flats. They may even consider a tree that is in a back garden if it can be seen from the road and adds to the visual quality of the area.

Some councils will only listen if you have reason to fear that the tree may be in danger of being removed or severely felled. Say there is an aged oak in a neighbour's front garden and the neighbour moves. You overhear the new occupant saying how much they hate trees and they are going to have it removed to tarmac the front garden and lay fake grass to put their noisy motorbikes on (shudder). You may be able to prevent this with a well-placed TPO.

If that doesn't work, there is a final level of action we could call 'being a little bit more than a little bit activist'. To protect a tree that an insurance company has said must be felled, or a local council has decided to remove, requires a mixture of charm and what we could call 'bloody mindedness'. You can't afford to make enemies of the council or, as one tree activist told me, 'Often the councils will cut down trees that the public tries to protect just to save face.' On the other hand, you need to generate publicity to shame them. But not too much shame: you need to be as cunning as an urban fox. It's hard and those who try to protect urban trees often fail. But sometimes they succeed.

*

In the corner of a field belonging to a UK primary school, Thundersley Primary School, in Essex, stands an oak tree that is 160 years old, and 50 foot high. The children adore the tree and the staff too. Not to mention all the tiny wee creatures that call the tree home.

In the spring of 2024 an insurance company told the school to cut down the tree because its roots may have caused cracking in the conservatory of a neighbouring property. The school would be taken to court and liable for damages if it refused to remove the tree. The head teacher asked if anyone was certain that it was the tree causing the damage (the tree was 12 metres from the building) and were told that to do a root survey would cost them £45,000.

The staff started a campaign to see if they could save the oak. The school children joined in, learning about the tree's ecological value and what was happening in the UK 160 years ago. Generations of former pupils who had sat under the tree helped. Petitions were organized and the local press arrived.

Then a local group of tree protectors called 'Save Chester' heard the story. They have arborists and tree officers, experts who came to see if they could save the tree.

They told me, 'The aim of our group is to persuade insurance companies that taking down a tree should be a last resort not a first action, just in case it's the roots of the tree causing problems. We did our own survey and were convinced that the roots of the tree were not the problem. We blasted the insurance company with all our findings, new evidence and the solution: a copper root blocker which would cost them about £3,000. It's not hard to put in a root blocker – you just dig a trench and put it in the ground.' So that's what they did. The insurance company paid for the blocker themselves. The campaign had only taken about 2 months.

'I don't know what made them decide to pay,' said the delighted head teacher, 'but maybe it was because it was the right thing to do? We were so delighted and were able to show our children that you can fight decisions that you don't think are right and have a good outcome.'

In Newark two 100-year-old Corsican pines and two flowering cherries were to be taken down by the council to create a children's burial ground and columbarium.

They had reckoned without a small group of local campaigners from a conservation group called, 'St George's Trust'.

An experienced campaigner, Sara Chadd, spotted the bats. She wrote to the local council. This is the beginning of her letter, which I liked so much I asked for permission to use an extract exactly as she wrote it. She included a map, which showed the flight paths of the bats.

> *We undertook a preliminary bat survey using electronic equipment. Bats are protected under the Wildlife & Countryside Act. There were hundreds of bat calls recorded over one hour in the cemetery including 196 calls identified as Common Pipistrelles, Soprano Pipistrelles, Brown Long-Eared Bats and Noctules.*
>
> *Newark Cemetery is an important potential bat roost and foraging area containing many veteran and mature trees. The two Black Pine Trees tabled to be felled under this planning application represent a 20 metre canopy circumference next to a healthy veteran Sessile Oak Tree about 150 years old where 84 of the Bat Calls were recorded. This Oak Tree would have its roots further disturbed by the proposed addition of pavers and tarmac for a car park. The root zone will be at least as wide as the*

> *natural canopy and building regulations advise against development within 12 metres of the root zone of a mature tree to prevent damage to both the tree and the proposed new surface.*
>
> *The pine trees appear to be healthy 50–60ft high trees, probably over 100 years old with naturally creviced bark ideal for bat hibernation roosts. They are within 20 metres of the building and given the high amount of bat activity at this point, it is the council's legal obligation to undertake a bat hibernation and maternity roost survey in the building in addition to in the surrounding trees before any planning decision is taken.*

Having summoned the bats so successfully to her aid, Sara went on to tell the council that to proceed with this work was tantamount to insulting Queen Elizabeth as this was taking place recently after her death.

> *At this extraordinary time of national mourning it is perhaps bad protocol to set a deadline for a public space such as Newark Cemetery and we believe the public would appreciate a delay on this discussion by reason of respect for the Monarch.*

You will not be surprised to hear that the trees that they campaigned to save are still standing.

It's not just the threat of removal, but even the threat of harm that can alert a dendrophile. A row of lush lime trees stand at the back of a development of flats where my local eagle-eyed tree wardens in Wandsworth noticed a planning application from an adjacent house owner to extend and build up to the boundary fence. Because the wardens live locally they noticed that no mention had been made of protecting the mature limes that stood adjacent to the properties.

'We spend a lot of time looking at planning applications,' they told me. 'Builders are not usually careful with trees. They would obviously have to excavate and the roots of the limes could be damaged.' So they made a few urgent phone calls and sent some emails that contained exclamation marks to ask for an emergency TPO.

In just three days preservation orders were placed on the trees, making it illegal to do any work which would damage these mature trees. They were on private land and the council agreed that they should be protected. The builders then had to work in a way that they knew the trees would be taken care of; they couldn't just start lopping off branches to put up a Port-a-loo.

Plans had to be re-written.

The limes are all still standing and healthy, so TPOs can do their job and approve the preservation of trees. If they hadn't been protected they may have been so badly damaged that they could have ended up having to be felled.

In Haringey, North London, there is a group of dendrophile activists calling themselves SHIFT (Stop Home Insurers Felling Trees). They have a website that tells you the best way to save a mature tree that is in danger in a variety of different situations. And they have had successes of their own. Four ancient oaks were in danger of being cut down on the demands of an insurance company, which was insisting that the local council would have to pay the bill for a subsidence claim if the council refused to remove the trees.

Meanwhile SHIFT collected 9,000 signatures and started camping out in the woods all night and all day, sometimes in two-hour shifts, to protect the oaks. In this case they also said that they wouldn't be happy with the cost of underpinning an adjacent property being passed to the council. So the local council was happy with them too.

'This action was life-changing for many people,' one of the campaigners told me. 'There were people from all walks of life who would never normally have met.' They had this common goal – so cross-generational friendships were made. People from different social, economic and political backgrounds listened to each other. Local press came and wrote about the 'Queen's Oak Camp'. The national press took up the story, running pictures of school children and pensioners camping under the oaks.

At first, the insurers were determined that the trees would 'have to go'. However, with sheer perseverance, the campaigners managed to convince the insurers to fund independent reports by structural engineers and arboriculturists who determined that tree removal would be neither necessary nor useful. As a result, the insurers agreed that underpinning of the property would be the only solution.

The ancient oaks are all still standing for the community to enjoy. And hopefully they will stand there for another 100 years.

It's not always easy being a tree protector. In London one young man slept high up in a 120-year-old plane tree (in a hammock with a safety rope) for 150 nights while the legal campaign between those who wanted to save the tree, the local council and the insurance company raged on.

Local councils like to cut down trees in the middle of the night, which is why those who are really determined to protect trees resort to sleeping in them. This is another example where the campaign was successful and the tree is still standing.

But even if campaigns fail and trees are cut down, it certainly gives councils pause for thought. Plymouth City Council became the focus of national rage when it cut down 112 trees in the middle of the night, ignoring a local campaign to save them. But now Ali, who ran the campaign says: 'They think twice. They wonder, if they want to take down a tree, "Will this cost a lot in legal fees?" "Is it worth it?" "Can we just work around the trees?" And other councils are far more careful too. They don't want to "do a Plymouth" as they know it's reputational suicide.'

So, Ali would argue, even if your campaign fails, in another sense you still win because you've let the council know how much people value mature trees, that campaigners often take councils to court and it can be a very serious and expensive mistake to get out a chainsaw.

Also if you want to mount a campaign, help is out there because all those who have run campaigns before – whether they successfully saved trees or not – are on a website called 'Canopy' wanting to help you.

If you're reading this in a country where no group like Canopy exists, then you can be 'a little bit activist' by forming one without even leaving your chair.

And may we each have a place to sit where, in one direction at least, we can see at least one mature tree. And may we each be a little more dendrophile.

ENERGY 6
Gas, Electricity and Solar Panels

About five years ago I moved my energy supply from British Gas to Good Energy. I hunted around at the time for a supplier that had a genuine commitment to renewables rather than offering just PR and greenwashing. Pictures of trees and vague promises weren't enough for me. I did my research conscientiously and eventually chose the company that I deduced had a genuine environmental policy. I signed up for dual fuel from Good Energy.

Since starting work on this project, the government's energy performance certificate (EPC) process, which you'll remember from earlier in this book, has left me wondering how best I can be more environmentally friendly in my energy use. So I'm going to go and see my energy provider in person. They have a famous founder and chief executive, a woman, Juliet Davenport OBE, who has kindly agreed to answer my questions.[12] I explained in emails that I'd been one of her customers for five years and I sent her, in advance, the tale of my EPC adventure which I was hoping she could shed some light on.

Juliet is so warm and down to earth that she offered to share her lunch with me while apologizing profusely for needing to combine my interview with eating. I'd guess Juliet is a similar age to me, immediately warm and quick to smile and laugh. To quote Juliet's X page, she has been 'Finding solutions for everyone tackling climate crisis with clean energy since 1999.' No greenwashing here then.

Good Energy's office is in Chippenham, so I'm enjoying the fresh air that always pleases my lungs on a trip out of London. We sit down in an attractive office with Ian, her assistant. Juliet seems delighted to take a break from hard work and chat to me.

I have to ask her: 'How did you end up as the CEO of a renewable energy company?'

'Erm …'

'Briefly.'

'OK. I didn't start my life as an environmentalist. My dad was a rally driver and my mother was an artist – there was nothing to do with the environment in my childhood. Until I went to university when I was 19 and studied atmospheric physics as part of my degree.'

'Very impressive.'

'To be honest the rest of my academic career hadn't been very impressive.'

'Sounds impressive to me.'

'I found atmospheric physics absolutely fascinating. When you study this subject you bump flat out into climate change. This was a moment when climate change was just starting to be written about in the Sunday papers and the science was very much there. By the time I left I knew I wasn't going to be a scientist cos that's quite full on. I spent a while moving around doing bits and pieces of jobs. I went and worked in the West Indies for six months – I'm not entirely sure why. Then I worked in PR for a bit and soon realized I wanted to do something else. I went and did a second degree, this time in economics, and then joined the European Commission on Energy Policy.'

'Sounds great.'

'And that was when I started to ask myself: if I was going to do anything about energy, what was I going to do? And having been pretty much technology agnostic when I went to Brussels, I came back convinced that renewables are a pretty good answer for a lot of things. Not just carbon but also security of supply – where we get our power from and what countries we give our money to when we buy our energy. I went to work for a consultancy and there was a German entrepreneur who was running an energy fund in Europe who told me that energy markets are opening up. He wanted to set up a supply company in the UK and asked if I'd do it. That's where I started.'

I applauded. 'Perfect.'

'I hope that was quite short.'

'My focus is on what we can do as individuals – our points of power as consumers. I know that we have one national grid that we all receive our energy from but by having you as my energy supplier, I'm putting renewable electricity into the grid. I realize that gas is another whole subject but I'm doing what I can as a consumer which is, first and foremost, to be with a reliable renewable energy supplier.'

'Yes.'

'I would have thought that the next most logical step would be to power my house with electricity so that I'm increasing the demand for electricity and hopefully, as renewables grow more in strength, taking renewable energy off the grid?'

'Yup.'

'But, as you know, when I had the government energy man round, he told me that he'd support me having as much gas as possible. At which point I was ready to believe that there is some kind of government conspiracy going on.'

Juliet laughed.

I ask desperately, 'Why the hell is the government not encouraging me to use electricity seeing as we have renewable energy and they have CO_2 emission guidelines to meet as soon as possible?'

She waited politely for me to stop.

'OK.' I smiled. 'You can speak now.'

'It's like this. Everyone's home is different. So each person has a different journey with this. There are different things that people can do depending on where they are and what they've got. I always think of a circle because you might start at a particular place on that circle and move round. It might be that the first and best thing that you can do is to change your supplier so that you're not giving your money to a company that supports fossil fuels. Or you might decide to do everything you can to improve your energy efficiency and change your supplier afterwards. Or you might want to generate your own power in your own home – we have a large group of households that do that – or you might start to change the way that you use energy.'

'In what way?'

'You might like to look at what times of day you use it. So you switch it on and off at different times of day to support the overall grid, which helps more renewables come up. If it's very windy or very sunny, do your washing. This is an easy example, as the supply of renewable energy will be plentiful at that time. There are four or five things that you can do in your own home.'

'But that doesn't explain why …'

'Hold on …'

I do interrupt. It's such an inexcusable habit.

'The trouble with the government is that they are being lobbied by the energy-efficiency lobby who insist that you must address energy efficiency first. They believe that this is the most important thing and they all nod and say, "Yes, yes – that's right." And it is right because it's the cheapest and most important thing to do because it reduces people's demand. The trouble is, that morphs into an argument by the energy-efficiency lobby that you shouldn't be able to count renewable energy as zero carbon. Otherwise there is no carbon incentive for people to reduce their energy use more.'

'But that's crazy.'

'The way they see it is that if you, as a household, take renewable electricity then you would go zero carbon without doing any energy efficiency at all – just by buying a renewable supply. They think that's a bad thing. So their system is designed to force you to consider your energy efficiency first.'

'So if, hypothetically, I could afford to change all the heating and cooking in my house to electric and had you as my supplier I'd count as zero carbon? But I'd still be taking the same electricity off the grid as everyone else. I wouldn't really be zero carbon.'

'But with us as your supplier you would count as zero carbon. In the business market you can say that. Under UN guidelines, if you are putting 100 per cent renewable onto the grid then you can claim that as your impact.'

'I thought only 50 per cent of the electricity we are using in the UK is zero carbon?'

'That's too high.'

'It's 33 per cent,' corrected Ian.[13]

'So, if I was 100 per cent electric with you as my supplier, I could claim to be zero carbon but I'd actually be using energy from non-renewable sources? So I wouldn't really be zero carbon?'

'Yes, but it depends on how you think about it.'

'As an individual consumer, I'm trying to use as much renewable energy as possible, increasing both supply and demand. The best way I can do that is to be with you isn't it?'

'Yes, that's right. And what the energy-efficiency guys are saying is that if you are with us but don't do any energy efficiency in your home you're still using more energy overall.'

'This much I have understood. They are measuring your energy overall, not the carbon you are using.'

'Yes.'

'And correct me if I'm wrong but I understand that it takes less energy to make gas than it does to make electricity?'

'It does historically. If you are running a gas power station, you put in a certain amount of energy and you only get 60 per cent of the energy out in electricity. But with wind you have infinite amounts of wind and you don't get that conversion piece. That's an old-fashioned way of looking at energy conversion.'

'What, saying that gas is more efficient?'

'Yes.'

'So why is the government still supporting gas so heavily?'

'Because they have supported the gas industry for years and years and it's a massive lobby. The construction industry has always had free gas connection, for example. If you want to build a new village, the gas industry comes along and will put in free pipelines for you. The gas industry knows they will make that money back over years, so for them it's an infrastructure investment. The government is happy and doesn't need to get involved.'

'But please ... surely I'm not a crazy conspiracy theorist, am I? Surely from the climate point of view aren't I right in thinking that it's insane that in a climate crisis the government is supporting the gas industry?'

'Yes. You're absolutely correct.'

'Thank you. I thought I was going crazy.'

'Take a look at Denmark. It's got 11 per cent of households connected to the gas grid. The UK has 80 per cent of homes connected to the gas grid.'

I try to get my head around this.

'Please eat your lunch, Juliet. I have an old gas boiler with an outlet that leads out of the house and spews out heaven knows what but it is certainly my own personal contribution to pollution and climate change. And the government inspector instead of telling me to get rid of it told me it was marvellous.'

She laughs at my frustrated rage. Obviously she's been dealing with this for years. I rant on ...

'There are hundreds of these officials going around the country to every single house that is rented or sold and encouraging everyone to use gas. Shouldn't they be encouraging everyone to use electricity?'

'Yes. But it's not how the EPC process has been written.'

'Isn't this a policy that needs to be changed as soon as possible?'

'We've managed to get this changed in business but we haven't managed to change it for domestic use.'

'When you say "we", that is the energy providers lobbying the government about the system?'

'Yes. But there are lots of competing voices. We have the fuel poverty lobby insisting that gas is cheaper so we need gas. You have the gas industry there and they are a very powerful lobby and they are working very hard to ensure that we are using the gas infrastructure because they are worried that it will become obsolete.'

'Well, hopefully.'

'You have all these different voices and when you have a new industry, as ours was, it's very difficult to get your voice heard because you don't exist.'

'But we have a climate emergency. We have a climate emergency! Even the government has declared that we have a climate emergency.'

'There is now. But this has been going on for the last 10 or 20 years. And this stuff takes years to push through.'

'I'm not mad then?'

'No, you're not.'

'That's a relief.'

I was beginning to wonder. I refer to my notes with some small renewed confidence. Ian smiles at us. Juliet eats. I find a new question.

'From the point of view of those who are also trying to do what they can to help the planet – assuming that they have a reliable renewable energy supplier – what else do they need to do? Should they be getting rid of the gas?'

'If they are able to get rid of gas that's great but not everyone can. I have an air source heat pump but I have outside space and they're expensive. I have solar panels. Everyone who has a roof should have solar panels in my personal view. I think we should cover every roof we can in the UK with solar panels. Because that would change everything.'

'It would be amazing.'

'I'm passionate about the fact that there is too much energy concentrated in the energy industry. There always has been and if you look worldwide that's why it's so difficult to change anything because the power is so concentrated. So, if we all take the power back a little bit ourselves by either using less or generating our own … It completely changes the dynamic between us, the customers and the industry, and that will bring about a shift.'

'That's exciting.'

'You've got to look at your own personal household. In my house, which is an old house, adding secondary glazing on all my windows absolutely transformed the house in terms of liveability.'

'In my home the instruction to add double glazing was the only thing that made sense on their list of suggestions. All the rest of it was nuts. They were suggesting cladding the outside of a Victorian terrace.'

'That's the "tick-box" approach. They suggested at mine that I should add another layer of insulation on the inside.'

'Me too.'

'They didn't consider that I already have exceptionally thick walls as it's a very old house.'

'But surely, as we have a climate emergency, one of the most important and easiest things to improve is this ridiculous test?'

'The issue is that it's managed by the regulator. It's paid for by the energy companies.'

'Sorry, what?'

'That's part of the issue. And it's a five-year programme. To change the next one we need to be lobbying now.'

'But we don't have five years to spare to make every house in the UK energy efficient.'

'This is true. They haven't really thought about what they are going to do going forwards, which is why the climate committee report at least said not to put in new stuff. They haven't even thought about how to change the rest of it. But they are looking at how we can green up gas.'

'Let's get back to what we can do as individuals. You would agree with me, then, that the only thing that made sense when they visited my house was to have double glazing?'

'Yes, and another thing that's really important, but which isn't part of the test, is to insulate your doors.'

'Your doors?'

'The secondary glazing company that I used also put a seal around our doors because there was so much of a draught. From the heating perspective draught is one of the worst things in the UK because it's not so often very hot or very cold. What we have more of is a lot of damp and a lot of wind and so with any house, any cold wind is always going to chill the house down – particularly in winter.'

'And thick curtains make sense, don't they?'

'They do, but make sure that you have secondary glazing or double glazing first and then you put the curtains up as well.'

'And we do facilitate people generating power in their own homes. We are the second biggest feed-in tariff administrator in the UK. In 2004 we started backing early adopters.'

'So people can call you and say, "I want a solar panel on my roof. Can you help?" And you'll say, "Yes."'

'We will. We don't install them[14] but we make sure that we buy the power and sort out all that part for them. We work with individuals but we also work with organizations that fund people who want to put solar on their roof. So we work with local councils, we work with housing associations to put solar into different people's households. That, for me, is one of the biggest shifts that someone can make very quickly and simply – if you have a roof.'

'Just one question on that. I heard a story recently that someone in Totnes with ten solar panels on their roof was trying to sell their house and there was some kind of problem as they didn't own the panels: they were owned by the energy company so it caused problems with the sale.'

'That shouldn't cause a problem at all. The lawyers should be able to sort that out. I'm guessing the problem was that they would have been put up for free and the rights to the feed-in tariff would be with the company that invested and the rights to the energy would stay with the house.'

'So, it's just an issue of legal paperwork?'

'Yes. Someone obviously hasn't done something correctly somewhere but the lawyers should be able to sort it out. We do changes of tenancy all the time for people with solar panels on their roof. I've just been through how many we do a week. It's a lot. If I was trying to decarbonize the UK, I'd start by putting a solar panel on every roof and then I'd look at heat pumps. What I love most about my heat pump is that the technology is so great.'

'So, if people do have money to invest, take a look at heat pumps? That's great. Thank you.'

'Next question?'

'Random question: my old gas hob has a problem. If I need to replace it should I get gas or electric?'

'That's a really good question. If I was going to use gas for anything, I'd use it for cooking but not for heating. First, because you don't use it a lot with cooking and, second, it's very variable so it's effective because it heats up very quickly and turns down very quickly. So I'd say if you don't cook a lot, choose gas and if you cook all the time, I'd go electric.'

'Hmmm. Well, I'll see if I can fix the one I have. OK, eat your lunch for a moment and I'll look at my next question.'

Juliet smiles and munches.

I find my next question.

'To frack or not to frack?'

'Not to frack. Gas has always been sold to us in the UK as a "transition fuel", cleaner than coal. Well, we've almost totally quit coal, and now is the time to transition. Not to extract more fossil fuels.'

'Nuclear power?'

'No, thanks. Too slow. Too expensive.'

'Too dangerous?'

'Yes, in the long term.'

In one of my rare moments of wisdom I decide not to debate the difference between the short-term and the long-term dangers of nuclear power. To my mind, it's immoral – another of those 'Let's just leave the waste for future generations to worry about.' But 'No, thanks' to nuclear power is what I want to hear.

'So … Moving on then. Smart meters. Love them or hate them?'

'Lots of people hate them.' Juliet says diplomatically.

'A friend who does not have you as a supplier told me that her provider literally harasses her to have a smart meter put in. They phone her and email her and text her. She blocks them and then they phone her again from different numbers.'

'We don't do anything like that.'

'I rang up a friend this morning who actively hates them and she told me every single evil thing that you ever heard about a smart meter – including that, according to her sources, they cause migraines, anxiety and tinnitus, and very specifically, if you have sperm in a petri dish and you put this in the range of a smart meter they start to swim backwards. Apparently, she's seen that on film.'

I swear I saw a slightly dubious smile flicker on her face.

'That's a new one. I haven't heard that one before. I'm not sure how to respond to that one particularly as I haven't done that experiment myself.'

'They say that the worst thing about them is that you can't switch them off. She is someone who switches off her Wi-Fi at night but she says you can't switch smart meters off and that she could write a short thesis on the evils of smart meters. But you like them?'

'There are two things with smart meters. There is something called a "home area network", which is where the meter communicates with something else which you then carry around the house. This is called an IHD or an In-Home Display unit. This is a thing that we have been lobbying against ever since smart meters were introduced.'

'You've been lobbying against them?'

'Yes, because they are a completely useless piece of junk.'

'What do you really think?'

'It's absurd. They are giving you yet one more piece of electronic equipment just so you can read the meter. Why do you need one of those? You have a phone and you have an internet connection. But they want us to give you one of these boxes because, according to their research, nobody is capable of using their phone. So we have been lobbying against this type of connection, which is constantly live. In other words, it has to be communicating all the time.'

'This is what my friend says.'

'Yes, she's right. But you can operate your smart meter in dumb mode which is where you switch off the IHD and put it in a drawer. This is what normally happens, which is why I think they are totally useless.'

'People don't even use them as they are designed to be used?'

'There is a battery in them and when the battery runs out people put them in a drawer and never use them again.'

'So is this what we call a "smart meter"?'

'No, this is just the display unit. The meter itself is separate. The meter goes where your old meter was – normally outside the house somewhere and nowhere near the main part of the house. As far as I've understood it, all the alleged health issues are related to the communication between the meter and the display unit.'

'But isn't it communicating with all the electrical appliances in your house as well? And that's how it's receiving its information? From your fridge, your washing machine and everything else in your home?'

'No. It doesn't do any of that.' She munches her celery.

'Doesn't it? Shows how much I have understood.' I laugh.

'No, it only talks to the IHD, which is usually in a drawer. A "smart" meter is not so smart.'

'Ha ha ha. I'll quote you on that.'

'Please do. All the suppliers need to do is record your data. The meter doesn't need to go into the house at all. We want the reading so we don't have to raise our CO_2 levels even more by having men in vans driving around, coming to your door, turning up when you're not there or trying to look at those ancient spinning meters that don't always work.'

'Ah yes, I've got one of those. Apparently, they work really well.'

Juliet laughs. I really like her. She's so good-natured.

'Sometimes they do.'

'That's what they told me when they came round. They said, "Those ancient meters – there are very few parts in them that can break."'

'Every seven years we have to replace all of them.'

'Really? Well I've had mine for about 25 years.'

'It needs replacing. You're probably on my list.'

'I'm sure I am. When your colleagues at Good Energy tried to replace my meter I refused to have a smart one. I have to say that the woman I spoke to was very patient with me when I refused to have one. I read later on that she is supposed to go to "all reasonable means" to persuade me to have one. And she did that but in a very good-natured way.'

'One of the biggest problems we have now is tracking down the old meters.'

'Yes, she said that. I sat in the dark for four days.'

Juliet laughs. 'I do think there is a going to be a market in buying up second-hand ones.'

'But, as it turned out it wasn't the meter – which was fine – it was a different problem. So I still have my old one. I'm willing to consider all reasonable offers for mine.'

'They are really, really difficult to find. We usually recommend smart meters in dumb mode.'

'She tried that. I still told her that I didn't want one. She was very good-humoured.'

'We know that there is a large percentage of our customers that don't want them. But seriously, smart meters are complex. Government research says that there isn't any health impact. We hear a lot of different things but we don't have any evidence that there is anything wrong with them. But we don't support IHDs anyway. What we want is a system where the information goes through your Wi-Fi directly to your phone. Everyone has Wi-Fi now.'

'So, the head of Good Energy says they are not smart and if you do have one just put it in dumb mode.'

'Yes. And we are currently testing a system that communicates via your Wi-Fi.'[15]

'What do you wish all your customers knew and would understand?'

'Our customers? They are quite well informed. We talk to them and they read our newsletters.'

'OK – so what do you wish the people who are not with Good Energy would understand?'

'What the difference is between us and a greenwashing company like Shell.'

'I heard a radio commercial recently. It started with a little girl's voice saying, 'I want to save the penguins.' It turned out to be an ad for Shell boasting that they can offer 100 per cent renewable electricity.'

'Oh, right?' Ian suddenly speaks, incredulous.

'"Yes – but you're fucking Shell," I yelled at the radio. I felt like phoning the radio station to complain. And are you committed to keeping oil in the ground? No.'

'There is a race going on between BP and Shell to try to prove who is the greenest.'

'I suppose it's a good race.'

'It's a good start, but the trouble is that to make this transition there is a lot of investment that needs to happen and it has to happen over time. What they're trying to claim is they have done it overnight. The majority of what is on their balance sheets is oil, not renewables. And if either of them genuinely want to be environmentally honest and call themselves ethical then they need to stop investing in oil. And all the other stuff they are both doing that is destroying the planet.'

'Quite. But enough about Shell and BP. It puts me in a bad mood. I prefer to talk about the companies that are doing it right. What are your aims as a company?'

'If you look at our purpose, it's not very catchy.'

'Not catchy is fine.'

'We talk about making a cleaner greener world with our customers. Energy is fundamental to society. We want that energy to be greener and cleaner. And when we say cleaner we mean enabling technologies like batteries, mobility and a whole range of other issues.'

'Good.'

'And we want to do it in a way that looks more like Airbnb than Trusthouse Forte. This means that it doesn't need to be us owning it all. We'd like our customers to own the infrastructure and we want to be the glue in the middle between customers that want to generate their own power, customers who want to be a part of the market, and to help customers who possibly want to change their demands so that they can get better deals during the day. We want to be the facilitator for all of this.'

'Is energy cheaper at night than it is during the day?'

'Yes – but after midnight. Energy is most expensive between 5pm and 7pm in the winter when everyone comes home from work, switches everything on and has a cup of tea.'

'I never knew that. So, if you have the option, don't do your washing during peak hours.'

'If you travel by train you probably know there is something called peak and off-peak but when people use energy they never think about this or the fact that it costs us more to generate power in the winter than it does in the summer.

People don't consider this. We've had quite a parent–child relationship with our energy providers for years.'

'That's interesting because we now trust our doctors far less and if a doctor gives us a diagnosis we say "Oh really?" and often go home and look it up on the internet. But people believe in their energy providers.'

'The providers have deliberately made it complicated. There is a lot of "belief" in the energy industry. When you describe people you can say, "He's a nuclear belief person" or "He's a gas belief person". And many people in the industry have this belief that their technology is primary, is right and is better than everyone else's. Anyone else who tries to get into this conversation is in danger of being overawed by all these "beliefs" that the *Spectator* or the *Daily Mail* write about. And these beliefs are often just based on views and opinions rather than reality. They are not based on a holistic understanding but on a specific piece of the overall picture.'

'Yes. I see.'

'When I talk to people to try to work out their logic, they can get very impassioned about their position. They also get impassioned about power cuts. There was one recently and the views expressed were as if the apocalypse had come. But there was a power cut and nobody died.'

'I was working on a fully charged computer all that afternoon and I don't put house lights on during the day as I like natural light so I didn't even know that we'd had a power cut. I think power cuts are a good thing as it encourages people to think more about power.'

'I agree. It makes people think about what a power cut is and where their power comes from. There is an edict from on high that we must never ever have a power cut. And that's ridiculous because if you never have a power cut, you build a system that is incredibly expensive.'

'I loved it when I lost my power for four days. We had evenings, candles and time. It was beautiful. It was a bit chilly. But we found woolly jumpers.'

'Yes. I'm not suggesting that we should switch our power off but we should definitely understand it better.'

'And break our habits.'

'And give people the power back. There is a bunch of lobbyists in the fuel poverty lobby representing people that can't pay their bills and they say that they can't afford to change their behaviour. And I said, "Have you asked them?" And then they did a piece of research that demonstrated that people, particularly those who are unemployed or based at home are quite happy to have a shower at a different time of day to save themselves money.'

'But they didn't know?'

'Exactly. And it's so patronizing to say that they must have cheap power at all times of day. Why not give people a choice?'

'So the times of day are when?'

'The most expensive times are between October and February and between 5pm and 7pm weekdays.'

'So, the best time to have a shower?'

'There is a peak in the morning between 7am and 8am. So if you're an early riser and are on a tight budget then have your shower at 6am. Or if you have a showers before you go to bed, have them after 9pm at night. You begin to build up an awareness. If it's really windy, have your shower then. If it's really sunny, have a shower then. Because these are the renewable technologies we have on our system. One good thing about smart meters, if we can do it without frying everybody …'

She smiles as she says this.

'… is that if it's super windy or super sunny we could send people a message to say, "This would be a really cheap day to do your washing." And give you that choice. So you get to the point where the house becomes an energy trader and so when it's windy you want them to use the energy.'

'How do I cut my CO_2 if I don't have a roof?'

'It's still about the time aspect that we've been discussing. The first thing to do is change your demand habits so that you don't shower or heat up your boiler between 5pm and 7pm.'

'I've a feeling those may be the exact times it's set to come on.'

'Yes, so change that. Everyone can do that. Then have your double glazing and your doors done. The double glazing company should be able do that.'

'What did they do on your doors?'

'They put on those plastic seals. You know the ones that if you try to do it yourself they always fall off?'

'Yes. My grandmother had them. They fell off.'

'If you get someone to do them properly they don't fall off.'

'What are they called? Just old-fashioned draught excluders?'

'Yes.'

'And presumably don't take down your doors? There was a time wasn't there when it was very fashionable to take all the doors down.'

'Doors are good and also you can have a heat block and put it in the middle of the house. A bit like an old-fashioned fire.'

'OK, last question. Aren't you, like all companies in this rampant capitalist system that we have, forced to make a greater and greater profit every year?'

'Have you read Kate Raworth's book *Doughnut Economics*?'

'Not yet. But you're not the first person to have recommended it.'

'She talks about whether a business is exploitative or regenerative. Although I hadn't thought about it in that way specifically, it turns out that Good Energy is up there on her regenerative list. It's all very well saying that if you spend more, then you're going to increase carbon. But don't spend money with companies that are exploitative.'

'Like tobacco companies.' (To pick an example entirely at random – bastards.)

'Just spend money with people who are doing the right things. One of the reasons I studied economics was so that I could argue with economists because they are very stuck in their mindsets and say things like, "That's an efficient economic process." And I'll say, "Yes it is, but it's not the outcome we're looking for, is it?"'

'But you still have to make a profit every year?'

'The interesting thing about that is that if you're making capital and putting it back into a regenerative process then, from my point of view, that's a good thing.'

'That's logical.'

'I had this great debate with myself, about ten years ago, about exactly that, and the reason for growing is to convert more of the country to low carbon technologies. That's why we're here. That's what we're going to do.'

'Good answer.'

'And we need fewer of these companies that exploit people and planet, and more companies like us. I'm not suggesting we need a larger energy market, we don't – we need a smaller energy market. But it needs more like us in it.'

'What are the other renewable energy companies apart from yourselves that you'd recommend?'

'The three that have got official OFGEM (Office of Gas and Electricity Markets) approval are us, Ecotricity and Green Energy.[16] We're all slightly different. Ecotricity is very much about the future: they still have to prove that they can do what they say they are going to do. Which I'm sure they will. Ours is based on where we are already. And Green Energy is an OK company too. They are smaller than us.'

'And Octopus?'

'Octopus I know quite well. As a group they made a big investment in solar. Octopus themselves are focused on giving great customer service. They do renewables but they do a 'light-green' version. They are nice. Of all the other companies, I like Octopus the most. They are a nice light-green wash and would be a good first step – but of course, as obviously I would say, with the ultimate aim of getting to the higher level of using Good Energy as your provider.'

'Well, obviously.'

Before I left we somehow got into a discussion about the fact that Juliet grows her own courgettes. I bet the CEOs of Shell and BP don't grow courgettes. That, I thought as I walked to catch my train back to London, is the sort of information you should know about the head of your energy company.

THE BIGGER PICTURE 3
Voting, Canvassing and Wolf Tattoos

As I'm doing everything I can to support the planet, this will include voting. I've never been loyal to one particular party. Some of you with a firm political stance will see this as a bad thing, but I've always looked at the way the country is, considered the most important issues and voted for the party that I think is dealing best with the social issues of the day.

I have voted for the Green Party – the party that has the environment at the centre of its political agenda – a few times. But, because I know that they stand no chance of being elected in my area, it has felt like a wasted vote and it has seemed as if it would always be that way, unless I move to Brighton Pavilion and into a safe Green Party seat.

However, the pledge of this book is to do anything and everything that I can to support the environment. This means that I can now stop tactical voting. Hooray. I'm done with all that. And so, truly joyfully and wholeheartedly, from now on I will be voting Green because I'll be voting for the planet.

If any other party comes along and says that, for them, the environment is the most important issue and ecological considerations need to come before economic ones, then I'll be able to look at the rest of their policies and make further decisions based on that. Until then, becoming an environmentalist, for me at any rate, will mean that every single other political consideration will come after the question of how we can take care of our planet.

Furthermore, there are many who believe that civil disobedience isn't a very enlightened idea. They claim that we have a perfectly good system for changing our leaders and it's called voting. But there is nothing to stop us from doing both.

I'm about to make my first ever phone call to a political party.

'Erm, hello, is that the Green Party? I was wondering about these next elections that are coming up … Do you need any help? It seems to me that there are a lot of undecided voters out there. And I'd like to tell them that I think we need to start putting environmental decisions first so I thought that maybe I could volunteer to do some canvassing for you?'

In less than an hour a member of my local Green Party arrives at the house with a large pile of flyers.

'Which streets would you like to do?' She shows me a map of where they had already covered.

'Maybe the area between Battersea Park Road and the park? On both sides?'

'That would be excellent. You're not nervous? Banging on strangers' doors on your own?'

Not until she asked.

'Greens don't usually encounter aggression on the doorsteps.'

What does she mean, 'usually'?

'It's OK to try to persuade people to change their vote, that's what canvassing is all about, but we just don't advocate arguing with people because trying to tell them they're wrong can be counterproductive.'

'OK, I'll attempt to persuade where I can but not argue. I was planning to listen to people.'

'Don't go anywhere after dark obviously.'

'I won't.'

'Don't take any unnecessary risks.'

'I won't – but why? What's the worst thing that's happened to you while you've been canvassing?'

She hesitates.

'OK, don't answer.'

'Most people are friendly. I'm sure you'll be fine. Keep a list of the streets that you cover. That would be very useful.'

'I will. You can give me more of those flyers. I'm planning on doing two full days. There's a lot to learn.'

She looks very happy. I suppose that for people who have been politically active all their lives, canvassing must be one of the least popular jobs. But for me – someone nosey and interested in everything and everyone – it's fascinating. And I'm following my heart too.

I put on jeans and a flowery blouse with a smart green jacket, hoping that this makes me identifiable, and start by using my immediate neighbour for practice – the one who played the piano for the Extinction Rebellion event. I ring his bell.

'Are you canvassing?' he asks, confused.

'I am. May I ask who you're voting for?'

'I'm undecided.'

'Well, I think looking after our environment is more important than all economic concerns.'

'Yes, I agree and we certainly need to change and prioritize that.'

'That is why I'm voting Green.'

'You're right. OK, I'll vote Green then. I'm tired of all the others anyway.'

I walk down the street and see a woman in the garden with her dog. I remind her that tomorrow is a polling day and ask who she's voting for.

'I'm undecided,' she says.

I tell her why I'm voting Green. She says that she agrees with everything I'm saying and she'll now vote Green. Great.

'Gosh, look at me go. I'm a natural,' I think as I walk happily away from my own patch into unfamiliar territory.

I ring the doorbell of a rather lovely house. A man answers, sees my 'Green Party' badge and scowls.

'I've already voted and I voted Liberal Democrat.'

'Wonderful. And may I ask the main reason that you chose that party, sir? I'm interested in why people are choosing whoever they are voting for.'

'I'm voting Lib Dem because I've always voted Lib Dem – it's inbred and ingrained.'

Inbred? I think to myself. Isn't inbreeding something that we should be avoiding? Interesting choice of words there. But, genetics and personal relations aside, personally I think that voting for a party because you've always voted for them is the wrong reason. Give me a policy that you particularly admire. Tell me that you're impressed by their manifesto, that you admire their leader or even that their values and priorities are most in line with your own. Just don't tell me that you vote for them 'because it's inbred' – because it's who your parents voted for. That doesn't impress me at all. But of course I said none of this. His tone doesn't invite discussion.

My very presence on his doorstep seems distasteful to him.

'Well, thank you. I'm glad you've voted.'

'Goodbye.'

I was in one of the wealthier local streets.

At the next house a small nervous face peeked out.

'My lady not at home – I only housekeeper. Sorry, English not good.'

I was sorry not to be able to talk with her or with 'my lady' but I gave her a flyer.

'I give my lady.' As if she doesn't count. I wonder where she comes from.

At the next house, with neglected paintwork, an angry man says, 'I'm not interested in any of it. Any of the politicians. They're all useless. I'm not impressed with the way any of them handle anything.'

I could have said, 'Yes, well, in that case you may like to vote for the environment and vote Green.' But he isn't asking for my opinion.

The next house has a profusion of breathtaking orange roses blooming in the front garden. I'm smelling them happily when a delightful elderly woman opens the door. She is proud to tell me, 'I'm voting Conservative because I've always voted Conservative. I paid £12,500 for this house in cash years ago. Now my home is worth over a million.'

'How lovely for you.' I smile at her pride in her house and in her chosen party. If only everyone had a home like this. It's wonderful that she knows how fortunate she has been, but isn't this a little unusual as a reason for voting for a party? Anyway, I'm happy for her.

The next house greets the visitor with an unfriendly digital intercom system instead of an old-fashioned doorbell. Another lady puts her head around the door without opening it: 'I'm sorry, I'm just the housekeeper.' She looks like she might be from the same part of the world as the last housekeeper.

The next house is another unfriendly buzzer. Some properties have no flowers, just concrete in front of their homes. Some actually have plastic hedging and plastic flowers. Really. Nothing to make those who live there feel glad that they have arrived home. I ring the buzzer – talking to intercoms is not meeting people.

'Hello. I'm here to remind you that tomorrow is voting day and, if you have a few minutes, to let you know why I'm voting Green.'

Click.

Hmmm, very friendly.

I ring the next identical expensive intercom and repeat my friendly speech.

'I'm not voting,' says a voice. Click.

Our democratic system is a miracle. Whether you live in a house worth over £1 million or in sheltered housing, whether you have a doctorate or no GCSEs, you still have one vote. One person, one vote. It's such a good idea. How has it all gone so wrong?

The complications of our 'first past the post' voting system as opposed to proportional representation befuddle my brain. I know it has been said that 'first past the post' tends to eliminate extremists but, on the other hand, it's possible for an unpopular candidate to win if the opposition is divided because the only criteria is who has the most votes. With proportional representation, all the votes matter as they are counted in a different way and seats are allocated proportionately. So proportional representation is a fairer system. But the idea of using PR seems to be a threat to our major parties. Some would argue that in a hung parliament, which proportional representation is more likely to create, it's even harder to get anything done than when one party has a majority. And so that's why we stick to a system that everyone knows is unfair. Beats me.

At the next house there is a woman in the front doorway. Two children who are evidently not hers hold onto her skirt bottom. I smile at her and point at my Green Party badge and then at my flyer.

'My lady is not here. Sir is also not here.'

Sir is not here? What is this? *Downton Abbey*?

'I'm here about voting. Surely you can vote?'

'No.'

I walk up to the next house with increasing trepidation. It's another anonymous buzzer but a doormat in the porch reads 'Welcome', so I ring and address the intercom system.

'I'm calling about voting day tomorrow.'

'Oh, wait there I'll come down.'

Phew. Friendliness is free after all. We chat. I repeat my main reason for wanting to vote Green.

'I was going to vote Lib Dem but your reasoning is sound, so I think I'll vote Green too.' Yayyyy – canvassing works.

At the next house a rather overbearing man, who looks at me with thinly veiled disgust, is just going out. 'Er, I was just going to say that tomorrow is voting day and can I tell you why I'm voting Green?' I said as he walked past me to his car.

'You have my wife's vote,' he said. 'She's voting Green. I'm certainly not voting Green. You don't want to ask what I'm voting because you won't like it any more than she does.'

I'm grateful that I'm not this man's wife.

The next house has rather exaggerated security with several security cameras. What are they afraid of? Domestic burglary in this part of London is at the same level as it was in the early 1980s. Burglars rarely bother as nothing has resale value. In my home the only things worth stealing are a laptop and a piano. My laptop is usually with me and the piano would be hard to carry out. If burglars came to my house I'd have to apologize that I had nothing worth their effort. But this house is advertising that it might be worth the trouble of breaking into. I press the bell, wondering whether the hallway will be lined with gold. A rather timid face opens the door. 'My lady not here.' What's is going on here? I swear to you she looks exactly like the last four housekeepers I've seen. Curiosity gets the better of me. 'Wait,' I don't want to frighten her but I find I really want to ask, 'You – where are you from?'

'Indonesia.' She says. 'Sorry my English not good.' Click. She shuts the door nervously.

I'm not sure how I feel about these wealthy houses importing maids. Isn't it fairer that we each do our own cleaning? I remember a scene from the film *Gandhi* where the Mahatma takes the tea service from the servant, thanks him and then serves his high-profile guests himself. And what about the environmental impact of wealthy UK citizens flying in domestic staff from Indonesia? What a world of extremes we live in.

I turn away from the most elegant streets to a slightly less absurdly priced road. I ring an old-fashioned brass doorbell and a young man aged about 25 answers.

'Yes, I'm voting Green because the environment is the biggest issue we have,' he says before I have a chance to speak. 'We're all voting Green in this house. Have you heard of the charity Population Matters?'

'Erm … yes.'

'We need to have less children.'

'Yes. We do.'

'The thing is, this is THE most important thing we need to do to help the planet.'

'Well, this is true but I often find it's people that already have a family or people that don't want children who talk about this one. Perhaps you think that's unfair but it often seems to be something that "other people" need to do differently rather than what we can do differently ourselves. Have you tried suggesting that your friends no longer take international flights? I mean, even for a year? It's a way to lose friends. So, not to have a family – it's OK for you to say this now but if you have a woman you love who wants two children …' He stops me.

'I'm only going to have one child.'

'Good for you. But what I was going to say is that if you love a woman very much who wants two children, you may feel differently. Obviously six children is irresponsible, even if you have enough money to give them a good home and a good education.'

'I think any more than one per couple is irresponsible. Look at this chart …' He runs into the house to get his computer. Well, this is refreshing.

'Look here: the human population has doubled in the last 100 years. It's going be almost impossible to save the planet anyway, but with between 3 and 4 billion more people it's going to be that much harder.'

'If it makes you feel better, I only have one child and I'm not planning on having any more.'

'Good, but you have to tell people. This is the most important thing anyone can do to help the planet. Have one less child.'

'Do you think anyone decides to have children or not to have children based on concern for the future of the planet?'

'If they really understand the issue, yes, I do. Just have a look at the website populationmatters.org. Really. All other environmental action is insignificant compared to this. You have to tell people this.'

'I promise. I will tell some people this. What's your name?'
'Brody.'
'Well, it's a pleasure to meet you, Brody.'
'Anyway, of course I'm voting Green. What other party is there? Bye.'
'Bye, Brody.'
Goodness. That was a bit different.

I ring the door of a friendly looking house opposite. The door is green at least.

'I'm voting Lib Dem,' says a well-built, big-haired woman looking conspicuously at her watch.

'Well, that's a great vote and I'd just like to say I'm voting Green because I think the environment is the most important issue.'

'My husband does the recycling, so that's OK.'

I contemplate saying that there is a bit more to do for the environment than recycling. But she doesn't look ready for me to ask. 'Have you considered going waste free?' She closes the door.

I move on to her neighbour's beautiful house with a shiny white door.

'I don't wish to tell you who I'm voting for.'

'Oh, OK. I'm just asking as I'm interested to know which policies are leading people's decision making.'

'I'm not saying.'

'Alright. Well, I'm glad you're voting. Have a good day.'

'Goodbye.'

Another fancy buzzer: 'I'm sorry. The lady isn't here. I only housemaid.' You have to be kidding me. She actually said 'housemaid'. Am I in some kind of time warp? I'm dying to say to her, 'Please tell me what your hourly rate is, what your hours are and whether you work six or seven days a week.' But I'd probably scare her to death. I do hope these women are treated well. No male housemaids?

So many properties are empty. Obviously owned, but empty. Do the owners live abroad? Are these second homes? I never realized there was so much empty property near me.

Often when someone answers a buzzer I get as far as, 'I'm here about the polling day tomorrow and I'm volunteering for the Green Party and …' They just hang up. One man says 'Go away.' I find this so strange. I look at the huge car outside his house. Maybe he works for the oil industry. I hope he's happy.

One house has a small Chinese flag in the front window. The woman who opens the door says, 'Yes, I vote because I'm a good citizen.' But she won't tell me who for.

Another lady becomes quite strange when I ask if she'd be happy to tell me which party she votes for. 'No, I won't tell you. I don't know who you are working for.'

'Working for?' I smile. 'You think I'm getting paid to do this? I'm a volunteer.'

'But I don't know that, do I? I don't know if that's true. I could be targeted.'

'Targeted? I'm sorry, I don't understand. Look – don't worry. Forget I even asked. Really. Have a good day. What beautiful roses you have in your garden.'

I positively bolted from her fear of me.

Next house. 'Are you voting tomorrow?'

Slam.

This job is hard. So the day goes on: I skip lunch, I keep going. But I can see why people bother with this because there are a good number of undecided voters who are happy to listen to the one reason I'm giving – that we have to put the planet first – and who decide that they agree with me as a result of our conversation. 'You'll have my vote,' they say.

One woman answers the door with a smile, sees my badge and says, 'I'm no good to you. I'm voting Labour.'

'And may I ask your main reason for choosing that party?' I ask.

'Yes, you may and I'm happy to tell you. I'm voting Labour because I've been homeless in the past and thanks to the Labour Party I had support from the state and now I'm living in this house. Without Labour policies I wouldn't have had that support. I vote for them because I think we have to prioritize addressing the profound social injustice in our society.'

'Thank you very much.'

I stopped to chat to this lady, telling her that, disappointingly, I haven't found a single Conservative voter who would give me any reason for voting Conservative other than, 'I've always done so.'

I moved to a poorer area. Here, for the first time, I saw leaflets for a far-right party that I don't even want to mention the name of. I found, much to my sorrow, that seven out of ten of the houses I spoke to in this area of social deprivation were voting for them.

Litter lies in front gardens. Bins are overturned. I hope it's the local foxes. All the doors have signs on them: 'Stop! No salesmen. No callers. No canvassers. No junk mail.' Almost all the letter boxes say 'No Flyers'. What is a volunteer Green canvasser to do? Do I leave a flyer or not?

'I will vote Green because Change UK didn't,' says one man.

'Sorry?'

'They didn't.'

'Oh. No. Er. Yes. Thanks for voting Green.'

On one street a man said, 'I'm voting for them.' He holds up the far-right flyer. 'Same as everyone else on this street.'

'Really?'

'Yes – I know them all and I know how they vote.'

I decide not to challenge him on his politics.

One man, whose muscles take up most of the doorway, informs me that he's an ex-con.

I talk to him about voting. He's planning to vote the same as his neighbours.

I point at some stunning wolf tattoos on his arm. 'What about these guys?'

'What you saying?'

'I'm saying that above and beyond all our human politics we've got to start caring for the planet first. I mean, in 100 years from now we'll all be dead, but the planet will still be here and we have to start making choices now that support the wolves, that support nature. That's the most important issue now. It's more important than all the politicians. That's why I vote Green because I care about nature and preserving it for future generations.'

'I see what you're saying.'

'The other parties put other considerations before the planet.'

'I don't like any of the politicians, really. I'm like you. I like nature. I care about the animals more than people.'

Spotting a picture of some dolphins hanging in his hallway, I go on: 'What about those guys? We have to stop the pollution of the oceans.'

'You're not wrong. OK, you're right. OK, I'll change my vote. I'll vote for you green people. I like that.'

Good grief. I've shifted a far-right voter.

I walk away very happy from that encounter. And it wasn't even that hard. But he's right. All his neighbours are voting far-right too – the ones that are at home. And most of them are not open to discussion. It wouldn't be hard to move them but the tabloids aren't giving space to the Greens any time soon.

I'm happy with the Labour voters that say, 'I'm voting Labour because there is no equality in education.'

Or, 'I'm voting Labour because austerity sucks. Tell me … when will it end? This austerity thing?'

'I, er, I have no idea. But I do know that the Green Party is anti-austerity.' Phew.

At the end of the day, I finally find a Conservative voter who says to me, 'I'm voting Conservative because I'm a conservationist and I think they can do more for the environment than the Greens can.' And I would love to stop and talk to her and ask her why. But she very politely excuses herself, saying that she's already voted.

And then it starts to get dark and I've only got five flyers left when my housemate rings to tell me she has made supper and I should come and eat it. So I walk home, bathe my aching feet, eat and tell my housemate these stories.

This may be a democratic way to change the system but it's a slow one. My day will have created a minimum of ten extra Green votes, I reckon. Maybe a maximum of 20. I didn't count. And some of the people I spoke to were voting Green already. I take some pleasure from the fact that all those who find a Green flyer in their letter box will at least know that the Greens were there. Maybe that counts for something too.

And I know that there is one man with wolf tattoos on his arm, who has served time for something, who will be drawing a cross in the Green box with his nature-loving arm. It may be 'important to move the environment up the political agenda', as I've been saying all day, and I hope that more votes for the Green Party will do that – whatever happens to politics in the next ten years – but, personally, as I fall asleep thinking of his wolf tattoos and how beautiful they were, knowing that this one man has changed his vote is enough for me today.

EARTH 4
Yippies and Wombles: Living off Grid

Can you imagine a life where you never again have to do a day's work in an office or touch a computer? A life free from advertising hoardings, where you don't have to be bombarded with instructions to buy something as soon as you walk onto the street? Where no one hands you free papers telling you how to think? Where you don't have to do work that you may not agree with? You don't have to commute or travel on an overcrowded train? You don't have gas, electricity or water bills? Where the only bill that you have to pay is your council tax – and even then you're on the lowest band? Have you even seriously imagined escaping the rat race?

As humans, we are using far more resources than the planet has. Apparently, we are using up the resources of one and a half planets. If we go on using energy as we currently do, we will need two planets by the mid-21st century.[17] But there are humans out there, in every country, who are living as if we only have one planet. Some call it One Planet Living. They take no energy from the national grid (in fact they put energy into the grid) but they have power from solar, wind and hydro, and some clever individuals even have gas that they create themselves using homemade 'biodigesters'.

In this quest to learn as much as possible about how we can live gently on the earth, this week I have come to Lammas Ecovillage in south-west Wales to take a short course and to meet people who have done this and others who are thinking of launching themselves into such a life. It's all quite extraordinary.

The government doesn't allow you to buy some land and just build a property on it. It's not that easy. In 1947 an act went through Parliament which stated that owning land did not automatically give people the right to build on it. You needed planning permission. What the government wanted was for people to live in the towns and cities and only the farmers – i.e. people that were actually making a living from the land – to live in the countryside. This protects the countryside. If we didn't have such an act, every rich banker who works in the city would buy land and build a home on it. It would be similar to the second homes disaster that ruins so many of our smaller towns. People buy a second property and visit for two weeks a year while property prices are driven up so much that young couples who are born and raised in the town can't afford to buy there, even if they are both working in full-time jobs. Many small towns have suffered greatly because of this. But our countryside remains protected from the invasion of the stressed city dwellers. Since 1947 the rules have become stricter still, but this, as it turns out, is a good thing.

If you are going to buy land and hope to build a property on it, you have to prove that you have a genuine intention to live off your plot. You have to build sustainably, and local authorities create a hundred hurdles that you have to overcome. Those who are not serious will give up and return to the city. This week I'm sharing the life of some who have succeeded.

About ten years ago, Tao and Hopi set out to find some land where they could build themselves a new life. They found a bare 'sheep-shagged' hillside. The sheep had eaten everything and the soil was of the same poor quality as most of the Welsh hills that are covered in the Mesopotamian ruminants. But the land, although windy and harsh in winter, was south facing and they had a water source. They bought the land and set about enriching the soil in every way, so that they could grow food. They planted an orchard. They planted trees for windbreakers and trees, plants and herbs known to enrich the soil. They resurrected an ancient and redundant hydro system that gave them fresh, drinkable spring water. They created hedges and they planted lots and lots of vegetables. In order to be granted planning permission they had to prove that they could sustain themselves and their two children from the land and make enough profit to have at least £3,000 per adult left over to cover their council tax. They also had to produce a detailed plan of exactly what they wanted to build and what they were going to build it with. There were enough building regulations to give a saint a migraine.

But they did something very clever. Instead of fighting the regulations and the local council, they befriended them.

'They are just human beings,' Tao said when telling us about the experience. 'They are overworked and stressed and they just need us to do certain things so that they can tick the boxes. As soon as they see that you are not trying to hoodwink them or do anything behind their backs, they're happy. We just kept asking what they needed and we learned to understand the regulations as well as they did so we knew not just the details of the rules but the spirit of them too, so that if there was a problem we could have an honest discussion with them. They love coming here now if we have a new building. It's so much more interesting for them than all their other work and they can see what we are achieving here. We are also helping Wales meet its carbon targets as our carbon footprint is so low. There was huge resistance to the idea of an ecovillage here when we first arrived – petitions and locals with 'Ddim Lammas' sheets in their windows. But now they have seen how we have enriched this place and they like us.'

The first day of our four-day visit is designed to give us some of the experience of living off the land. You could choose to work in the polytunnel, learn to scythe, harvest fruit, replace the lime wash on one of the buildings, separate and repot new plants, or weed around one of the ponds where one plant was dominating at the expense of others. I notice they are sparing us the more difficult jobs. 'Clean out the compost toilets' and 'scrub the goats' shed floor' aren't included.

First we are shown around. Hopi talks with obvious pride about the building of her home. 'This house cost us £12,000.' It's a simple enough home. We are shown into a large kitchen and dining room with steps that lead up to a bedroom. 'The children's rooms are out the back.' It's a really lovely space.

'We designed the building around the windows because we went to a builder's yard and they had seven double-glazed windows that were the wrong size for the building they had been made for. We asked how much the builder wanted and he gave them to us for £5 each as he was so glad to see the back of them. So then we designed the house around them. As you can see, they are square windows but they are plastered around the edges to make them look rounded. It's an easy trick.'

There is nothing I have that they lack, except a flushing toilet. Someone asks about that.

'We have a good-quality compost toilet. It's one of the strangest things about the way that most people usually live. Our most valuable commodity – water – we shit and urinate into. And it's even water that's been cleaned. Here we have more respect for water than that. Our water is all drinkable spring water so we are much happier with a well-maintained compost toilet.'

'How did you get on with all the building regulations?' asks a participant who has already told the group that he plans to go down this path.

'We made lots of mistakes at first. We tried to take shortcuts and not stick to the rules. We were taken to court and that wasted years. So we learned the hard way. Now we know the rules and we keep to them, so everything is much quicker. If we want to create a new building we have the men from the planning department up here immediately and they are involved every step of the way. It saves a lot of time and aggravation.'

'Does this work?'

'It does. The rules are all there for a reason. You see this staircase up to the bedroom? This wouldn't pass the latest building regulations as it's not safe. And – to be fair – if you try going up that ladder when you're half asleep, there is a strong likelihood of you injuring yourself. The regulations are there for the protection of the public, not to make the lives of the planning inspectors harder.'

'What are the walls insulated with?'

'Sheep's wool. There's a lot of it in Wales.'

'And the timber?'

'It's all from sustainable local woodland. You have to build sustainably. That's in the regulations too. We use a lot that is recycled. We'll go to a local builder's yard and see what is there and buy whatever we can for reuse. Almost nothing here was bought new.'

'And a shower?'

'That's in the goat shed outside. The shower is hot and lovely but you need to pull the shower curtain if you don't want the goats watching you.'

I rather like the idea of Showers with Goats as a Native American name. It doesn't have the romance of Dances with Wolves, but there is an everyday pragmatism to it that I could enjoy.

Tea is served in a large elegant teepee that amusingly has a five-star rating from the local hygiene inspectors as a location for making and serving food to the public. They have two teepees at Lammas: one for use as a kitchen and one that they bring a fire-bowl inside and use as a meeting room.

We chose jobs for the rest of the morning, being instructed to choose what pleased us most. I pick blackcurrants. This is especially wonderful because, of course, they are all organic and Tao says that anyone picking fruit can eat as much as they like. The sun is shining and it's all very beautiful – like a TV commercial for alternative lifestyles. I talk to a fellow participant who has come with her husband and a gorgeous baby boy of seven months.

'We just know we want to give him a more authentic childhood. One where he'll be in touch with nature and be able to play outside. We don't want to bring him up in a city. And we're not afraid of hard work so this is definitely an option for us. We have a little money saved – so we're exploring this very seriously.'

I chat to another woman who is on her own but determined. 'I don't care if I start in a caravan. All I want is a water source and some land. I just can't do the same job for the next 30 years. But as I'm single I know I'll need a community, which is why I'm thinking of a plot somewhere near here. So there will be help if I need it. I might come here to volunteer for a while so I can learn more before making any major decisions. It's a long journey and I have so much to learn.'

I pick blackcurrants. It's like picking blackberries but without the thorns. I take a turn carrying the baby around. I show him the flowers, the giant poppies, the marigolds. Two large ducks waddle up and down the rows of planted onions, picking out the slugs for their breakfast. Ah, so that's why my garden at home has so many slugs – I don't have ducks. The peace is so restorative to the heart of a battered city dweller. There is no sound from a car, even in the distance; just the leaves blowing a little in the breeze. We work our way down the row of blackcurrant bushes until Hopi shouts 'Lunch.'

They try to feed everyone with food from the land and, although they admit to cheating when they run these courses (there are 20 of us on this experience week), there is still a huge pile of freshly picked salad leaves, herbs and edible flowers for our lunch. Nothing in a plastic packet here.

In the afternoon I choose to go in a group to visit another plot that has been more recently purchased. Keith, who is on his own, bought a piece of land only three years ago and so, although his trees are very young, his vegetable plot is in full swing and he is living in a caravan while building his house, for which he has successfully obtained planning permission. He shows five of us around, patiently answering questions from those who are planning to self-build. They have questions about everything from the building materials to the rafters. His free-standing solar panels are wonders of engineering and beauty and he explains for the more scientifically minded how they store energy into batteries and make it available for him to draw on when he wants it.

Keith also has a large and impressive pond in which he hopes to keep fish that he will be able to eat. Currently the pond is deep but lacks one of the key components that you need for a pond: water.

'We've tried everything to get the water to stay in but it just drains away. First we used puddling clay, which is a traditional pond liner. The pond did fill up when it rained but then just drained away again. Then we used 2 tons of bentonite granules. That should have worked but it didn't. So for a third try we're going to use bentonite liner. That will work but it's going to cost £800, which is a lot when you're earning your money from growing vegetables.'

I choose a task from the many on offer: repotting elderflower plants.

'What will you make from the elderflower bushes when they have flowers and berries?' I ask. 'Oh, elderflower juice, elderflower jam – maybe even elderflower wine.' Two huge dogs bound up. 'I'm so sorry – they are overly friendly, I'm afraid.'

'That's OK. I love dogs.'

The rest of my group wanders off to various tasks. I'm very happy to sit on my own all afternoon concentrating on carefully separating the plant roots and repotting. In the evening we sit in the teepee around the fire. (Yes, this is possible – the smoke is drawn upward and vanishes through the space in the top, just like a chimney.) Someone plays the guitar while others sing and play drums.

They call this 'future living'. They don't see this as the way people used to live but as the way people will have to live in the future. And they see themselves as the trailblazers. More and more people come every year to learn about living in this way. If this is so, then I have seen the future and it involves living simply in co-operation with those around you, rather than in competition with them. And the woodsmoke smells good.

The rules that the Welsh government sets (there are different rules in every country) mean that you have to prove that you can make £3,000 a year per person from the land to pay your council tax, as Tao and Hopi told us, and also that you can produce a minimum of 30 per cent and a maximum of 65 per cent of your food from the land. All the information about how you are going to do this has to be included in your planning application, and the planners want to be sure that you know what you are doing before they allow you to move your family on to a Welsh hillside.

On the second day we are introduced to a range of different businesses that have successfully gone through this process and we're given the opportunity to question them on the details. There are many different ways to make money from land but today the examples are a cut flower business, a couple who keep sheep and chickens to sell meat and eggs, a man with an ancient non-interference method of bee keeping who sells raw honey, a man who creates the mycorrhizal fungi that helps trees establish their root networks, and a business making natural cosmetics using plants and herbs. Lammas itself has an orchard, amongst other things, and residents bottle and sell their own apple juice – but that's seasonal.

One project strikes me as particularly brilliant. A very clever businesswoman explains, 'We thought it was too much to buy the land and build a house and start a business that has to succeed all at the same time. So we decided we'd start the business and get that up and running first and then buy the land and build the house.'

So she started her project, which was to make and bottle an innovative range of kimchi. Kimchi is a staple of Korean cuisine. It's salted and fermented vegetables flavoured in a variety of ways to taste. It's delicious and very good for the gut and her products are now successfully selling through local health food stores and other outlets.

'At the moment we are buying the food locally. But the idea is that as our own vegetables start to grow we will use more and more of our own homegrown food, which is chemical free, until eventually we will be using all our own produce and making our living entirely from the land.'

We listen in admiration. Her jars of kimchi are here for us to taste so we all bite into the tangy, salty vegetables and predict that they will continue to sell well.

'We don't need to make a fortune. We just need to make enough to meet the requirements for living from the land. And as I enjoy cooking, this suits me very well. We don't price the product too high as we want it to be enjoyed regularly by the local community.'

The man with the bees is equally fascinating. He explains that he's using a method of bee keeping developed by a monk in France in the 1820s.

'We started with a philosophical decision that we weren't going to do bee keeping in the intensive way that is taught on most bee-keeping courses. This is low-intensity bee keeping. We don't interfere with the bees. Some modern methods of bee keeping use plastic hives and they open the box once a week – which the bees find highly stressful. With this ancient, proven and very simple method, we haven't treated the bees for anything. We believe that if people have to use antibiotics on their bees then they are doing it wrong. The hive itself is an organism and the bees themselves organize it. It's not just a collection of individual bees. People wonder why bees are dying. It's not just the chemicals. It's the way we treat them. And this honey has a qualitatively different taste.'

He is selling it for £4 a jar.

'This is way too cheap,' I complained. 'This is like the caviar of honey. How many bee flights with tiny little bits of honey on their legs, from your own carefully nurtured wild flowers, has this taken? This is crazy. This should be £20 a jar. At least.'

He smiles appreciatively. 'You're right but we want to sell it to the locals. And they can buy cheap honey for £2 a jar in the local supermarket.'

'But there is no comparison. We don't compare wine that is under £5 a bottle with wine that is £20 a bottle. We know that there is a significant difference in the quality. So why isn't it the same with honey?'

'You tell me. It just isn't.'

'I tell you what – when you sell it . . .' I climb onto my high horse with my unsolicited advice '. . . get a jar of that cheap honey and then get a jar of yours and let the customers taste the difference.'

'That's a good idea actually.'

I feel like a bee protector. He is selling beeswax lip balm for £2 and honey still on the honeycomb for £4. We try some of the honeycomb. It's unlike anything I'd eaten before as it's raw – untreated – and so you can see and taste the pollen in it, still powdery in sections. And it doesn't taste waxy, like a commercially bought honeycomb, just sweet. It doesn't last long. We devour it, making a range of sounds of delight and satisfaction.

Then, though it isn't officially part of the 'business talk', a very clever former vet shares his thoughts with us.

'Working as a vet I was picking up on the damage caused by intensive farming. I was a livestock vet and I just couldn't do it any more. I could see the causes of the problems I was there to resolve and I couldn't do anything about them. It's not a system that I wanted to have anything more to do with. So my family and I are doing this. We are living in a horsebox and four trailers at the moment waiting for planning permission to build a simple straw-bale bungalow. We've never been happier.'

Then he shows us his biodigester, although this too isn't officially part of the 'business' part of our tour. This is how he produces gas for his family. It's basically an artificial cow's stomach with a collection of bugs inside that break down foodstuff in the same way that food is broken down in the gut. He simply feeds in food and out comes a rich fertilizer for the ground: an antibiotic-free cow poo. It also produces methane gas, which he stores and burns. This gas isn't ideally something that the planet wants more of, but in this case he has a tiny pipe which connects to a second-hand gas hob from an old caravan. With this his family cooks. Eventually they will also have electricity from solar power and hydro – but for now they are using food and plant waste as a means of cooking.

I listen in amazement: the details of how this works go way over my head while other people around me ask intelligent questions. People are so clever. Hopi says that living like this makes people more creative because, instead of just turning up at a shop and buying what you want, you work out how to make it using whatever you have around or whatever you can reuse or recycle. Once again I realize how little understanding I have of the gas in my own kitchen, of how it is made and where it comes from. How disconnected I am – how disconnected many of us are – from life's vital skills. I'm not sure I could even grow lettuce. I promise myself that next year I will at least create a space – somehow, somewhere in my London home – and grow some tomatoes. Surely, at least, I can do that?

We walk back across the field for another homegrown lunch. The salad is filled with edible flowers. The conversations are also a relief to the soul.

'Can we disconnect ourselves from the system that we don't want to be a part of?' someone is asking.

'We have to. We have to link our economy with nature. We can't keep consuming from our planet at the rate we are doing.'

It feels so good to be surrounded by people who all feel the same. It's the same as when I was with Trees for Life and everyone was talking passionately about trees.

Someone asks Hopi about clothes. 'We buy all our clothes second-hand,' she says, flaunting a low-cut top that displays her cleavage rather magnificently. 'There are certainly enough clothes to go around.' There is a feeling of freedom here and of life being more real. 'We each are guided by what we love to do. Tao is passionate about the land in a very practical way. My first passion is human beings and how we share love with each other. I'm also very interested in shamanism. That's my spiritual path and that fits very well with the land.'

Later, people who live locally tell us about the ancient practice of coppicing the trees that doesn't kill the trees but provides firewood, about keeping sheep (which have to go to abattoirs to be killed) and, best of all, about how one of the locals has planted a field of walnuts. Apparently, these trees should produce up to 2 tons of walnuts a year, which they will be able to sell. It will take ten years before they produce a crop like that but no one here seems to be in a hurry. You have to think long term in this way of life. I suppose ten years is not so long to wait: if I think back ten years of my own life it seems like the blink of an eye. And I certainly didn't plant walnut trees.

I'm reminded again of that Chinese saying, 'When is the best time to plant a tree?'

Answer: 'Twenty years ago.'

'When is the next best time to plant a tree?'

'Today.'

I don't often wish I was someone else or at a different time in my life. But I find myself feeling a certain envy for the young couples that live here. If I were 25 or 35 and here with a partner with a huge brain who had several muscular brothers, I'd move here in the blink of an eye. I wouldn't mind cleaning out the compost toilets. Even if the young couples that are launching themselves into this life fail and return to the cities, they will have learned so much, lived in nature and not spent ten years in jobs that they don't enjoy.

I share these thoughts with a fellow participant.

'Yes,' she said. 'Doing a job you hate. Destroying the planet and buying shit to make yourself feel good.'

Hmmm. A little harsh perhaps. But it can't be denied that this is how many of us city dwellers live.

'This is why mental health is so poor: people having to stay in jobs they don't enjoy to pay their mortgage and not seeing a way out.'

'Well, some people have jobs that they love,' I suggest radically.

'I would argue that it's a more genuine joy when it's rooted to the land, when it's connected to animals and nature. There is certainly lots of evidence to show us that nature is good for our mental health. People turn to nature instinctively for consolation. After loss, for example.'

'Yes. I've seen that. I worry about my neighbours that never use our large communal garden in Battersea.' I think of them. 'There is a little boy. He's about seven. He will never go in the garden without his shoes on. He's afraid of the "dirt". And he doesn't like the neighbour's cats. His mother says that she couldn't identify any of the birds in the garden. I've tried to teach her but she's not interested.'

'And that little boy at least has a garden. We are raising a generation of children that are totally cut off from nature.'

'But not the kids here. Even the teenagers like it here.'

'Yes, that is extraordinary.'

'I'm going to study for two years before I even begin to consider this life,' my new friend says. 'Then I'm going to buy a plot that has a water source and a caravan and work it all out from there.'

'You're going to do all this alone? Without a partner?'

'I am. I'll come here for a while as a volunteer, immerse myself in this lifestyle and learn from the mistakes others have made. I feel really clear in my path forward.'

'Will you come here to Wales?'

'I'm not sure. There are lots of ways of doing this. Lots of different styles of ecovillages. I'll explore them all. I'm not in a rush.'

'That's one thing we are learning here. Real change happens slowly.'

We walk back across the land past all the new planting. And even though the sun isn't shining and the sky is a thick Welsh cloudy grey, everything feels just right.

Today we have a talk from Tao. You have to admire a man who started with a bare hillside and has created a community where nine families now live permanently and many others come and go every year to learn about living off grid.

Tao looks, perhaps, exactly how you'd expect someone called Tao to look like – a classic hippy. His hair is long, his face gentle, his skin is tanned and he has the deep penetrating eyes of a visionary. You might expect a man like this to be a little waffly or lacking focus and precision in his talk, but no flies settle on this man. He's been talking to people about creating this life for years. He knows his subject, has nothing to prove and smiles easily.

He starts in classic hippy territory by talking about dreams and visions. 'As he speaks though I realize that this is the same clarity of focus that some motivational speaker would advocate in some high powered leadership seminar that teaches clear intention.'

'To create the future you want you need to have a clear vision. Consider what is most important to you. Do you want to be near the sea? Is your relationship to woodland important to you? Do you want to be high up on a hillside with a view or nestling in a valley? Is there a lake in your dream vision? Who are your neighbours? Would you like there to be people on the land next to yours or would you prefer cows or trees? It's important to enjoy what the Native Americans call "dreaming intensely". And to acknowledge the importance of the dream.'

This is what other disciplines would call 'making a clear plan,' but it's a step that many people miss. If you find some land and buy without matching it to what you've planned in your head, you could end up at the seaside when you'd rather be nestled on the hillside or milking a cow when you'd rather be walking on the beach.

I've always been irritated by the idea of people 'manifesting' what they want and the whole 'divine ordering' concept because it so clearly doesn't work or we wouldn't have refugee camps full of families that just wanted to create a better life for themselves. Everyone would be living somewhere that they love. All actors would get the leads they want and all professional athletes would win gold.

On the other hand, the way Tao talks about it – it makes complete sense. He's saying, unequivocally, before you even consider buying any land, be very clear about what you want and how you'd like it to feel.

Then he goes on to talk about the structure of the vision and the new life that many of my fellows here are going to create for themselves.

'There are rules that underpin any society,' he says.

'You can step away from mainstream rules, but any society you join or create will also have rules. Even if you decide to live in a trailer and be a hermit, you will still have to pay your council tax. Wherever you live and however you live you will be, to an extent, dependent on your neighbours. You'll need to run a pipe across their land or need help to add a roof.'

This is why, at Lammas, they have people who are living this life alongside each other. So that when they need help they can call on each other.

'When we first came here we weren't intending this place to be just for our family. We were planning an ecovillage for many families. Local people sent over 1,000 letters of objection to the local council, but over time we won them round, so much so that we recently received a letter of apology signed by a large group. You have to be aware of the society you are moving into. Here, for example, they especially appreciate it when we learn Welsh.'

Fair enough. I can't imagine moving to Italy and not learning Italian. It would be rude. And the Welsh are fighting to save their language and their culture. Why would they want a bunch of people living on their beloved land if they don't demonstrate that they will love the land and respect the culture?

'The families that live here: all the children attend local schools where Welsh is compulsory. So they all speak fluently. We try to keep up with them.'

How wonderful.

'Different communities create different structures for their communities. Some are run on Gandhian principles. Some are a darker shade of green and some a lighter shade. We're a fairly light shade here.'

Yes. Gandhi was a vegetarian and I can see plenty of cars around here. They're light-green all right. But they are a more authentic shade of green than I am. So I'd better shut up and listen.

'The third step is finding and buying land. This is complex and takes time, so don't expect it to be easy. The rules for purchasing land are different in Scotland, England, Ireland and Wales. At the moment it's easiest in Wales and in Pembrokeshire because the local authorities now understand that we are a force for good: we're paying our taxes and helping them meet their carbon goals. In other places it's harder.'

'How do you go about buying land here?' the father of the baby asks.

'Word of mouth is the main route. We tend to hear about plots of land that are available near us. And then there are specialist local estate agents that deal with land sale.'

'What are the main things to consider?' asks the woman who is considering embarking on this lifestyle as a single woman. Kudos to her.

'The two main things you need are a water source and land that is south facing. The higher up you are, the cleaner the water is if you are near the source and the cleaner the air is. The quality of the soil is a vital consideration too.'

'So how long does it usually take to find land?' asks an older woman who was there with her new partner.

'I'd expect it to take between one and six months.'

There is a good mixture here of excitement and realism. There is an optimistic feeling – that this way of life can be achieved – and yet, living here, we can see that it's not easy. If someone forgets to add sawdust the flies soon find the compost toilets and someone has to clean up the mess. Their hydro power is off as they haven't had enough rain and Hopi has stopped keeping cows because she grew weary of being a milkmaid. The requirement not only to feed yourself but to make enough money from the land to pay your bills is a constant issue. This is not a life for someone who wants to lie in a hammock all day and watch the clouds go by.

Tao explains that the next part of the process is the planning – dealing with the bureaucratic process. And he goes over again the importance of having a very clear plan about how you are going to make a living and using the details of the local authority demands to aid your process, rather than letting it drive you crazy.

'You choose a plot and write a very detailed management plan about how you're going to live there in a low-impact way. The One Planet Council enables you to download and study every single successful management plan, so you can see how it needs to look and the sort of details that the planners are expecting.'

He holds up a successful application and passes it round. It's like a thesis in length and detail.

'Assembly of a planning application can take up to six months. It may seem like a lot of work but they actually want you to succeed. They don't like all the land being sold off for sheep farming. The sheep strip the land of all the biodiversity. We enrich the land. They can see that.'

'Isn't £3,000 per year per adult hard to achieve?' asks our youngest participant.

'That's why you need a clear plan. It's also to help you. Your goals are firstly to live sustainably, secondly to build naturally and thirdly to be productively working with the land in a way that benefits both the biodiversity and the humans.'

'And this is possible?'

'As you can see, yes it is. But not everyone makes it. When you're in the early stages, living in a caravan in December and it's pouring with rain and you've had an argument with someone you love, it's easy to ask, "What am I doing here?" and give up. So you need a clear vision. The planners can help if you remember they are human beings and not suits trying to make your life harder with unnecessary rules.'

'It looks like lots of paperwork. I thought one of the benefits of this life would be escaping paperwork,' says a man who works in an office.

'Yes. Lots of paperwork. You need to produce an annual report showing how you are meeting the criteria and will continue to do so. They want to see ecological improvement.'

'Good,' says someone. 'So do we.'

'So we recommend getting an ecological survey when you start. It's between £300 and £400 but it's your baseline so you can show how you have improved the land. This isn't hard if the land has been previously sheep-shaven.'

Ha ha ha ha.

'You need to demonstrate carefully that you know what you are doing. A soil survey is also a good idea. The Royal Horticultural Society offer a survey online that's only about £35 and well worth it so that you can show how you will improve the soil.

'The last stage is "manifestation" or making it happen. And the success of that depends very much on how carefully you've planned in stages one to four.'

'How much land would I need?' someone asks.

Everyone wants to ask something. Tao gives up his planned talk to answer questions.

'Two acres is the smallest One Planet Living plot – really only suitable for mushrooms, strawberries or cut flowers. Five acres is average and 7 acres is a good amount to make a livelihood.

'And how much would all this cost? To put a home of some kind on the land?'

'How long is a piece of string?'

'At the moment you can do the minimum with £25,000. £50,000 would be a better sum to start with and for £75,000 you can look at some of the best options.'

'Can you move onto the land before applying for planning permission?'

'I strongly advise against this. It was possible once but new planning laws here have made it very difficult and, most importantly, they don't like it.'

'And what's it like in England?'

'I'm not an expert on England. But there is a book called *The Rural Planning Handbook* by Simon Fairlie that is an excellent resource for English laws and guidance.'

'Are the rules as complex in England?' a man sighs.

'They are complex everywhere. They want you to present your plans in a way that they can accept you. So, learn the rules wherever you are and then play them to your advantage. In a sense the application is a work of fiction since it's a future projection and you can't really know how you'll get on. But as far as possible you want it to be truthful and real.'

'So, no one can just buy a field and a caravan and live in a field?'

'Not for more than 28 days a year. The aristocracy don't want the peasants to have their own land.'

'And you can't make your money virtually?'

'No. To live on the land and not in a city you have to feed yourself and make at least half of your living through land-based produce.'

'And how many One Planet Living applications succeed?'

'About 95 per cent get through. Not all of them. But we work very hard.'

'I don't like petty authorities,' says someone

'They are just human beings,' says Tao. 'But if they sense that you are trying get around the rules then they can make your life hell.'

'It's not for the fainthearted, is it?'

'No, it's certainly not. It's an opportunity that both humans and the planet need for us to recover our relationship with the natural world. And we still have to tick 100 per cent of the local authority boxes.'

'Do you think we could really do this?' asks the wife of one of the couples. 'We want to change the way we live totally.'

'Well, Hopi and I have done it and we help others to learn from our mistakes. So, yes, I think this is the future. It has to be. We can't go on living in the way we are. Humans are not happy and the planet can't sustain it.'

We are silent for a while. Taking all this in. Then Tao says, 'OK. Time for lunch.'

In the afternoon we get to learn about more of the activities we can enjoy here. I swing a scythe for the first time in my life and find that, like all skills, it's not as easy as it looks. But, this news is just out, you can clear a field of long grass without power tools. Without even so much as a rechargeable battery. All you need is a group of friends, a number of scythes, a sunny afternoon and tea. After a few hours of this I feel fantastic. In cities people pay money to join gyms. I run around Battersea Park listening to podcasts for no other reason than to stay fit. We have all become insane.

In the second session of the day I have a turn sitting on a traditional wood-shaving horse. It's a bench that you sit on which holds a piece of wood in place while you carve it with a shaving tool. I'm offered a piece of cherry wood and an hour later I have a cherry-wood pan scraper for my kitchen. Now I know that this is hardly the achievement of the decade, but I haven't made anything out of wood since I was about nine years old and 'whittled' a pointed stick with a penknife. So this is progress.

In the evening, we sit in the teepee with a bonfire in the middle. Hopi creates a visualization with us, much to the surprise of some of the more conventional participants.

She gives us paper and crayons and while she sings and bangs her drum, we are encouraged, as children would, to see a vision of how we would like to live and to draw it. An image of the life we would like to create – with pastels. We are asked to close our eyes and 'see an image' and then to make it into a kind of reality using paper.

This puts me in an interesting dilemma as I have not come here with an intention to buy land and change my life – only to learn about the process, to enhance my appreciation for those that do live in this way and to report on it for you in case you want to escape the rat race and are drawn to this way of living to care for our planet.

I close my eyes and can see only a large wooden gate. It's maybe the entrance to a field or something. I have no idea what is on the other side. So I draw the gate. I like drawing.

The girl next to me is clearer. I can see a waterfall appearing in the top right-hand side of her drawing; then a small house; then loads of vegetables, plants and trees; then birds and animals; and then two black cats appear. This is good. She now knows that if she ever has her home and her land she will have two cats on it. This is the way this magic works. You dream it. You draw it. You look at your picture and then, of course, you're more likely to create it because you've visualized very clearly what you want.

I draw flowers around my gate. Purple flowers growing on the bushes. It's not a new life but it pleases me. I'm not going to create this gate so I'll just have to see if I come across it. If I decide to move from my current home and I'm house hunting and see this gate … I'll know that I want the property behind it before I've even seen the building. Hopi is a little bit shamanka.

That night they make a fire and musicians arrive with guitars and drums. Wine and other alcoholic beverages appear from local shops. I order French organic red. I would have ordered wine made in Britain but they don't stock that in the local store[18] I would have enjoyed drinking elderflower wine made from their land, but after two glasses of French wine I was quite able to appreciate the decades of expertise from France.

We dance, we sing. We look at the stars and the moon coming up. And we are together, we humans, under the huge sky.

The following morning I stuff my sleeping bag into its little sack and vacate the tent they had lent me. I wasn't sorry to see the back of the compost toilets. They weren't smelly – they were just, well, full. Compost toilets are very clean and produce no effluent (apparently they had to explain this to the local council), but this week, with the influx of 20 more humans than normal, whoever had been allocated to filling up the sawdust to make sure that the flies stayed away had been neglecting their duties. Someone was going to have to deal with this when we all left. I was glad it wasn't me but I didn't like the symbolism of leaving waste of any kind for someone else to deal with. They would do well to have more compost toilets and fewer people on the courses. But they do a pretty amazing job.

The 'closing circle' is around last night's fire. Predictably, just as we are all leaving the sun has broken through the grey Welsh sky and it's a beautiful warm morning. We sit and look at the faces which we hadn't known four days ago.

A Welshman speaks first. 'I came to prove to myself that you're just a bunch of hippies. And you're not. And that's confusing. Now I need to go home and think about this seriously. I'll break it down into baby steps.'

One of the younger men adds, 'For me, being here has been about freedom and connection to people. I have so much appreciated being with people who care about nature.'

'We get high-quality, heart-centred people here,' Tao smiled.

A woman spoke. 'I've been reading a novel that talks about the fact that it's not the difficulties that we have in our lives that define us – it's what you do with them. Here you are quite literally creating life and abundance from something that was barren. That's beautiful.'

A couple who worked for many years for a conservation charity have also been part of the group. 'It's one thing working for the land,' the man said, 'but we can take connection to the land to a deeper level. It's been great to see all the ways that it can be done.'

Another man said, 'It's been like a return to my childhood. A reminder that everything is sacred. Being outside reminds me of the miracle of life. I want more of this in my life. My city life is all wrong. I don't know how I'm going to do this. But I have the vision.'

A burned-out activist spoke next. 'Once you have moved through the grief of the reality of climate change, you start to look at what you need to see and be.'

The woman next to him nods agreement. 'If I put my mind to one side, there is a strong pull from my heart and body to do this. So, if I can get my mind, which wants to control everything, out of the way, I can perhaps allow things to unfold in a more gentle way. I'm thinking about "relating to the earth as a conscious being" and it changes everything.'

Everyone seems happy. Living in nature for a week is certainly good for the soul.

Another man speaks. 'I have a deep yearning to be on the land. I've felt it for a long time. I've been thinking about what you told us about Goat Willow, that it can spring up from the smallest piece. It's encouraging.'

One man was on a return visit. 'I have always lived minimally and I got involved with the Transition Towns movement and the permaculture movement. I first came here eight years ago and, to be honest, I half expected it all to have just collapsed and folded. I'm hugely inspired to see that it hasn't and that you've found new ways, as a community, to move forwards and learn.'

A younger woman: 'I've been depressed and lost. The way my husband and I live just doesn't work for us. And thinking about the future as we had seen it: work hard, buy a house, work harder, buy a larger house – we know now that we don't want any of that. We want the planet to benefit from us being here. We want to be part of the solution. This life may not be for everyone but we are one couple that just can't live in the future the way that we have been living.'

One guy just said, 'I was losing hope but I see that there are still real people around.'

Another woman laughed. 'It sounds crazy. I had a tarot reading. The woman looked at my cards and said, "Sell your house, buy a camper van and go west." I decided to take her advice. I've done the first two. So maybe I'll come here as a volunteer for a while and then I will have done all three. I'll work out the path while living it.'

The last woman said, 'I'm so excited by what I've seen. And I'm very nervous, but my soul is already packing her bag and is halfway down the road.'

And me, you ask? I felt a bit of a fraud because I've just been here to learn about this life rather than genuinely intending to live like this. I know my limits now and, just as I didn't chain myself under Waterloo Bridge during the Extinction Rebellion actions but banged my drum to support those who were doing so, I'm not about to buy a field – but I certainly intend to keep in touch with and support those who are. Brave planet-loving souls.

Back in London, much to my surprise, a friend is very critical of this way of living. 'They're running away. It's cowardly. I have a friend who is working in a failing school. For International Book Day he took his class to a bookshop because none of them had ever seen or been inside a bookshop before. That's the level of deprivation he's working with. He's my hero – not those Wicker Man yippies.'

'I'm sorry – "yippies"? That's not a term I'm familiar with.'

'Extreme hippies. Did they teach you to use a crossbow? Admit it – they did, didn't they?'

'Disappointingly, perhaps, no. Only a scythe.'

'Isabel, they are eco-woks. They can't really believe that what they are doing is a solution. It's an indulgence. That way of life isn't open to everyone. It's so privileged.'

'It's more affordable than trying to buy a house.'

'So is everything. If the ability to buy a house is the benchmark of inclusion, we're fucked … and it's copping out.'

For a moment I heard 'it's coppicing out'. Maybe I've become more yippie than I've realized. I'm not sure my London friend would understand the work involved in coppicing. Maybe he thinks the yippies just grow pot and sleep till lunchtime.

'It's hard work making a living from the land. It's not an easy lifestyle choice.'

'It's ridiculous. They're a bunch of Wombles who think they can save the world.'

Wombles find things left lying around and reuse them. They'd love to be called Wombles.

My friend doesn't seem to like them.

'It's make-your-own-catsuit-from-seaweed nonsense. And, most importantly, it doesn't answer how we can actually stop climate change.'

Goodness. Why would anyone be so enraged by people living like this? And he hasn't even met them. I wonder how many other people, who perhaps have never so much as planted a potato in their lives, think like this?

For the record, I'm not advocating One Planet Living as a life choice for everyone. But I certainly wouldn't recommend working for 40 years doing a job you don't enjoy to pay off a mortgage, or many other aspects of our rampant runaway capitalist way of living, either. Would you?

If I ever move I'd want a tiny house but with lots of land. I think my absolute favourite would be a large space where someone had put down concrete or plastic grass. And I'd take it all up and plant fruit and nut-bearing trees, and vegetables, and wild flowers for the insects and the birds. And I'd want my own roof so I could have solar power and, if there was enough space, I'd love a little wind power generator too. And space for a washing line where I can dry my clothes with a free old-fashioned mixture of solar and wind power.

For now, though, I have no intention of abandoning Battersea Park Road and life in this polluted and overcrowded city. Not for the foreseeable future anyway. There is a meeting tonight where my local council is discussing how they can make Wandsworth in London a zero emissions council by 2050.

The local Friends of the Earth team, the borough's Green Party and the closest Extinction Rebellion group, are all full of energetic people that also laugh a lot, drink red wine, make banners, shout and support the planet. We may not dance around fires, but we make a lot of noise in the public gallery at council meetings. We let the council know that we are passionate for change. Wherever you are, whatever you are doing and however you're doing it – there's a lot to do.

CLOTHES AND FASHION 1
Stylish, Sassy and Sustainable

Today my jobs include a little darning. Or at least sewing up a few holes in an old knitted cardigan with wool. The new wool, that I've bought in a local haberdashery, which opened in 1947, is as close a match as I could find to the original colour. Is darning just mending holes? I'm not sure what the difference is between these two activities. Which tells you how much I know about mending clothes.

My beloved woollen cardigan is over 30 years old. You could be forgiven for thinking that I can't have worn it much. But that's not so. Some years, in colder winters, I've worn it almost every day. It's a rich dark-beige colour that matches all my clothes and it's made of a thick wool that gives the cardigan a texture that looks great in photos. I can wear it over a top of any colour with a pair of jeans or I can put it on over a smart dress if I want to look a bit more casual and homely. I've worn it to go camping, to practise archery and to curl up on a sofa and read a novel. I've always washed it lovingly by hand, in tepid water with a minimum of soft soap flakes. Apart from the few small holes I am fixing today it's in as good a condition as it was the first time I wore it. If an edict went out that we were each allowed to have only ten pieces of clothing, it would be the first item I would choose.

And as I sit darning, thinking about the longevity of my relationship with this item of clothing, I notice the care with which I'm doing the task. The concentration that I'm giving to sewing up a few small holes is just a tiny reflection of the dedication with which the garment was made. This cardigan was knitted by hand and made just for me. I remember asking for it to be created. My friend was wearing a cardigan that was similar, which I complimented her on. When she replied that she had knitted it herself I was so impressed that I asked if she could make me one. So she did. I have no idea how long it took her – all I know is that it has a label in the back: Hand Made by Vicky. What makes this garment unique among all the other clothes that I own, is that it contains love. In each and every stitch.

To the famous question asked by environmentalists, 'Who made your clothes?' this is the one item I have where I can answer, 'a friend called Vicky Wicks who now lives in Bristol and has two sons at university. When she made this for me her sons were not yet born and she had yet to go to university herself or meet the man who would become the father of her sons.'

So you can see why this is my favourite item of clothing.

I remember when my daughter was four, a friend offered her the chance to be a bridesmaid. She was allowed to choose her own fabric and she stood excitedly on a chair while they measured her. A delighted and happy child then watched as my friend's mother cut the material and, sitting with her sewing machine, created the miracle of a little dress, with a matching belt, and even a matching hair-tie. That outfit was worn by that young bridesmaid with such joy that I can still see her pride in the photos. She knew the time and care that the adults around her had given to make her outfit perfect for her. Mothers and fathers reading this: quickly stopping by a store and grabbing a few new things for your kids – made by who knows who, who knows where – just doesn't create this particular happiness.

When I visited the indigenous Asháninka tribal people in the Amazonian rain forest and had the privilege of living with them for a few weeks, I watched the women spinning cotton they grew themselves and dying it with plants, and then weaving cloth to make clothes for the men they loved. All the men wear garments that their wives have made for them and they have observed the labour of their creation. Their clothes are also worn with pride.

So many of us in our developed world suffer from mental illness, and loneliness has been called the modern epidemic. I wonder how much better we'd all feel if at least some of us were able to wear clothes made by people we love. And I know that some feminists will be thinking that this would be a step backwards because it's mainly women who make clothes. But I'm not speaking of enforced labour, I'm speaking to those that do have the luxury of choice, can consider these questions and are seeking to express love – both women and men.

Our clothes are one thin layer between our vulnerable skin and the world. How much better would we feel if we were kept warm and protected by love in the form of clothes?

Apart from my hand-knitted cardigan I have a few items that have been altered by a local tailor, Mohammed, from Syria.

Mohammed sews very well and speaks three languages but, sadly for me, English isn't one of them so further acquaintance is challenging. I would have welcomed the opportunity to know a little about him. How he learned to sew, how long ago he left Syria and how he has ended up on Battersea Park Road. After all, his attention to detail mends my second skin. Aside from these few exceptions, though, I have no idea whose efforts are protecting me from the elements. I do know that most of what I wear may have been made on the other side of our planet, probably in Bangladesh, by women who work a 16-hour day, 6½ days a week and don't earn a living wage. They are unlikely to have time to use their skills at making clothes either for themselves or anyone they love.

Also I don't know the environmental cost to the planet of the clothes I have. Many of them are cotton. Some are linen and, before I look, I can guess that there will be a fair amount of polyester and other forms of what is basically plastic. I'm learning to be an environmentalist so all this is going to have to change.

With the exception of the Vicky cardigan, my clothes don't love me and often I don't love them. I'm not saying I hate my clothes. I tend to spend a bit more to get an item that will last longer. I've never been a follower of fashion, preferring more classical clothes. I look just like everyone else: I wear blue jeans, too much black – all the usual lazy clothing habits developed by so many of us. I've shopped neither for ethics nor for joy. I've just wandered in and out of clothes shops, a victim to having my attention fall exactly where the shop owners want it to fall, and often leaving with items that would later prove to be a mistake. My entire approach to clothing needs an overhaul. Speaking to friends, I find I'm not alone.

I'm sure I don't have to tell you that the fashion industry is the third most environmentally damaging industry on the planet after oil and animal agriculture. Apparently it has five times more carbon output than air travel. Cotton, which we all love to wear, is the fourth largest pesticide-consuming crop. The chemicals used to dye our clothes pollute fresh-water supplies. Shockingly, 90 per cent of the dyes that are used to give our clothes colour are dyes that are extremely damaging. And, of course, if you buy synthetic fabrics, they are made from petrochemicals and then coloured with yet more chemicals. Tanning of leather is also one of the most environmentally damaging industries, and that's without even considering the way the poor animals are likely to have been treated. I remember years ago the moment I realized the reason I loved leather. It's because it feels like skin and skin is good to touch. Some penny dropped in me: it *is* skin. It's the skin of some innocent large-eyed cow. I stopped buying leather.

What a mess we are in. Walking around in the 'developed' world, making ourselves feel good with the short-term dopamine hits of our 'shopping therapy'. As forms of therapy go, this is not only one of the most environmentally damaging but also one of the least effective. As high-street clothes have become cheaper and cheaper, owning them sparks less and less joy in us and we have all ended up owning way too many clothes. In 'developed' countries we buy three times more clothes than we did in the sixties – and yet many people feel they have 'nothing to wear'.

Greenpeace estimates that 80 billion garments are produced every year. Of those, on average three out of four of these garments will be incinerated or will end up in landfill somewhere in the world. A fractional amount of this is recycled: less than 1 per cent. Of the clothes that we keep, on average 30 per cent of what is in our wardrobes hasn't been worn for over two years. The Ellen MacArthur Foundation found that more than half of all 'fast fashion' clothes are thrown away within a year – that's 100 billion units of clothing. This is all contributing 1.2 billion tons of greenhouse gases per year to the atmosphere. And all for clothes that people don't really like – let alone love.

We've been tricked again. We are sold cheap clothes which often trap the people who are making them and pollute the planet that we live on, while the aspects of clothing that could give us the most joy – creativity and human connection – have been taken from us.

You may be thinking, 'Ah, but I take my used clothes to charity shops and they raise money for charity.' And this is true – to an extent. I have read that of clothes that are taken to charity shops, only 10 per cent of them are resold to the public in the developed world. I don't want to discuss the figures – no doubt in some charity shops it's considerably higher than that. In others it may be lower. What I do know is that anything with a stain or a tear or anything that those who work in the shops thinks that their customers won't want, leaves the shop to begin a long journey. They are moved to sorting centres and from there are put into large bails, then onto lorries, then onto ships, where – you've guessed it – they are shipped across the world to end up in second-hand clothes shops in developing countries. This has killed the domestic clothing industry in many countries.

'But at least someone will end up wearing them?' you say. Let's hope so. But the sad truth is that, just like our recycling, most of our clothes end up on open landfill sites. Having harmed our environment at their creation they then harm it again as they rot. The more you learn about the life cycle of your clothes, the less you want to be a part of this industry. We can help our purses and buy less. Way less.

On the other hand … I don't know about you but I want to have way more fun with my clothes. And clothing is one of the most wonderful areas to play in. And need not cost much. In fact, you can save a whole lot of money. Remember, I'm advocating that we get radical. So here is my suggestion, which is very simple: apart from underwear and clothes for those rare special occasions, which I'll take delight in discussing later, I'm suggesting that we rarely buy new clothes any more. And instead we really learn to enjoy and to celebrate our clothes. This may be considered a frivolous pursuit so let's fully embrace that idea. I'm happy to be a wonderfully frivolous ethical environmentalist. Frankly, the movement has room for this.

I've invented a new version of the Marie Kondo process for our clothes. I've found that her 'does it spark joy?' question can be too widely interpreted: there is a danger that it can be widened to include such considerations as 'It did when I could fit into it', 'I'm sure it will when I can fit into it again', 'No, but it cost a fortune so I'm keeping it anyway', 'No, but I like the colour', 'No, but I've only worn it once so I have to wear it before I can give it away.' And other miscellaneous excuses. I appreciate her weird habit of thanking an item for the service it has provided, but it's the 'giving it away' part that I'm increasingly uncomfortable with.

Ethical fashion activist Lucy Siegal has suggested that we don't buy anything that we won't wear at least 30 times. That figure seems low to me. But there are still items in my wardrobe that wouldn't pass it. So, let's take a look in our wardrobes: I have a number of new games to suggest.

My new variation of Marie Kondo's process is to open your wardrobe and take out all the items that you'd award a score of 10/10. Just the items that you really love to wear. The ones that make you happy every time you put them on. Not the 8/10 ones. The 10/10 ones. Imagine that you could only keep, say, ten items of clothing: which are the ten that you'd take out first? Take those out and lay them on the bed. I'm not limiting you to ten, but I'm asking you to take out the ones that you love the most – the ones you feel happy wearing. Your favourites. Don't say you don't have any favourites – you can't feel that bad about your clothes. If you do, then discuss this with a few wise friends. They'll help you find ways to improve this.

Assuming that you now have your favourites on the bed, do a spot check of the items remaining and remove any that you know need to be given away for any reason: you've never worn them; you're never going to wear them. Ideally, share them with friends but if they are going in a charity shop pile, they'll need to be washed, mended and ironed – or they will go straight in the rubbish pile at the charity shop.

Now take out all the remaining clothes and put them somewhere else. Maybe into a spare wardrobe if you have one, or into a large storage box if you don't. They'll all need to be washed, mended, folded or hung up conscientiously (not crushed or creased). These are your give-away clothes for your friends. 'Hold on!' you cry, rushing to retrieve a couple of items for yourself, 'But these are my clothes.' Ah, yes, but they are not the clothes you love the most and life is too short to wear clothes that you don't love.

I'm suggesting, as a lesson in non-attachment if you like, that we all have a small cupboard or box of really good clothes that we are prepared to give away. When friends come round we'd all be able to say, 'That's my give-away cupboard, have a look through and help yourself to anything you'd like.' This way our clothes won't end up in landfill in China or Africa but will enhance the lives of our friends. We might own a shirt that we rate 9/10 and never wear, but it may become a 10/10 item to our friend and actually be worn with love. For clothes to be shared like this they will need to be good quality. From now on, we're simply not buying fast fashion or poor-quality clothes. So we'll need to think way more carefully about every single item we purchase.

Now take the 10/10 clothes, the ones that you really love, and put only those clothes back into your personal wardrobe. Hopefully there are not too many of them. Study them carefully. Mend anything that needs fixing. Sew up the little holes. Take any beloved items that need to be dry cleaned to the cleaner's. (Hopefully you have an environmental dry cleaner's locally that are using minimal chemicals? If not, ask them about this.) Maybe even take an item to your local tailor to have it taken in or let out or altered. Then make an inventory of exactly what you have. Having done all this I am experiencing a strange new feeling, something that I've never felt in my entire life: I look forward to getting dressed in the morning. I'm creating strange new combinations of my favourite clothes. I feel better when I'm out because I'm wearing clothes I love and, surprisingly, because I stop to think about what goes with what instead of just pulling on some jeans and some old top, I've started to look like someone who cares about how she looks. Apparently it's sometimes known as 'style'. This, believe me, is a very unexpected side effect of this project.

There is another game I know that you might like to play. They say that we wear 20 per cent of our clothes 80 per cent of the time, so if you don't like the idea of decluttering or giving away clothes that are not your favourites, here's a different game. Take out all your clothes on their hangers and then put them back on the rail with the hanger hook the wrong way round. The inconvenient way. Then when you wear an item and return it to the rail, place the hanger the right way round. This way you'll be able to see instantly what you've worn and what you haven't worn. Force yourself to consider the unworn clothes first. Give them a chance. If you make a note of the date that you did the multiple-reverse-hanger-turn then a year from now any clothes that still haven't been worn should be feeling nervous. Maybe then you'll be ready to give them to friends.

If you have trouble with clothes or know you have way too many and don't know where to start, here is another game to play with a friend. And maybe share a bottle of organic French wine if you're that way inclined. I recently asked a friend if she had an hour to help me look through my clothes and I tried on items I rarely wore.

'Why don't I wear this?' I asked as she sat on my bed sipping wine. 'I really like it off the hook – and I love the colour.'

'Because it looks terrible on you,' she offered scathingly. 'It goes in where you go out and goes out where you go in. It's not in the least bit flattering. You need to give it to someone who is a different shape.'

I gave it away that very day. She was similarly honest about the rest of my clothes. I was so grateful to her. Less, in my experience, is always more. It's just easier.

Where clothes are concerned, men are often naturally better at minimalism than women are but sometimes men also end up with 30 shirts too many. Why own rugby shirts that you don't like?

And, guys, please consider what you wear and how you feel in your clothes. I have an ex-boyfriend who, when I met him, had a collection of silk ties that had been given to him by the wife of a man who had committed suicide in tragic circumstances. Of course he thought of him every time he put on one of the ties. I gently suggested that the ties might be enjoyed more fully by men who didn't know the history of the previous owner.

We can also wear clothes to please those we love. Life is short. Why wear something that our loved ones don't like? Don't misunderstand me. I'm not talking about coercive control and telling other people what to wear. I'm assuming a loving relationship where, if you have three shirts you love and your lover likes two of them, you are able to dress both to please yourself and their eye too. Looking good for those we love is part of understanding attraction and is hugely underrated. Everyone dresses carefully during courtship but can we extend that attention to detail for our lovers over a lifetime? And we can enjoy dressing up for children and friends too. Don't wear jeans and a black top every single day. Life is too short not to celebrate with more colour than this. Parents, wear clothes your kids love, for fun. And there is even the ultimate test: wearing clothes that your teenagers approve of. This can also be a game. Ask your teenagers, 'Which of my clothes do you like best?' After all, there must be one or two items they will admit to enjoying seeing you in. There are many ways to have lots more fun with our clothes.

Many of us have become unconscious consumers. So you can discuss this with friends. If you were designing your clothes from scratch – for example, after a house fire in which everything you owned was burned – how many pairs of jeans would you choose to own? Three? How many T-shirts?

And so on. I'm guessing that you may already have enough clothes to last you for quite a few years. But let's imagine that during this process you discover that you don't own a light jacket or you don't own a cashmere sweater to keep you super warm and cosy in the winter, or that you'd really love some tops with more colour. Now you're ready to hit the charity shops. I'm recommending knowing exactly what you'd like more of in your 'favourites' wardrobe. Then start looking around your local charity shops as often as pleases you. The old trick of going to the charity shops in the most expensive areas works well. It's amazing what people who have way too much money will get rid of without a second thought.

Personally, I love to look for silk and cashmere in charity shops. These luxury fabrics just feel so good on our skin. I wouldn't feel comfortable buying these items new but I've often found clothes that must have cost an impressive amount of money new being practically given away for under £10.

If you are one of those people who has never bought second-hand clothing in a charity shop, you may be cringing at this point. For some people the idea of buying second-hand clothes is a step too far. But if this is you, please have another look as you'll find these shops have changed. Ten years ago charity shops were often dirty, and the clothes were tatty and smelled bad. Now – even though some are better than others and there are still way too many clothes on a rail – they look more like vintage clothing stores and everything is clean. Anything dirty, stained or needing mending they will discard. Everything is steam-cleaned and pressed right there in the shop. The clothes are more varied and individual than in the high-street shops so you can really play and experiment. And if you find something you like, you're not giving money to the fashion industry: you are making a charity donation, someone is giving you an item of clothing and you're rescuing cloth from landfill.

Shopping in charity shops (or 'thrifting' as they call it in the USA) takes more time than high-street shopping, but if you do it quickly and often (knowing what you are looking for), you soon develop an eye for what you really like and what you don't; you're able to play with style and really enjoy the frivolity of it all. That's one of the reasons I love charity-shop clothes. You can't take yourself too seriously. You know the expression, 'Blessed are we who laugh at ourselves as we will never cease to be amused'? That's what charity-shop clothes shopping is all about. The notion that status is about what you wear makes people who wear charity-shop clothes laugh.

How did we ever buy into the idea that to be 'someone' we have to wear designer clothing? Or something new? What absurdity is this? Audrey Hepburn said, "The beauty of a woman is not in the clothes she wears, the figure that she carries, or the way she combs her hair. The beauty of a woman is seen in her eyes and grows with passing years.' Charity-shop clothing celebrates this. Both men and women can express who they are by enjoying who they are, not by needing to express it with expensive clothing choices.

I have a local charity shop that gives money to people whose lives have been negatively affected by AIDS. I found a shirt in there last week. It's a deep, rich red. It's made from some wonderful cotton and is apparently from some famous designer that I've never heard of. The details are a celebration of shirt wearing: cuffs that double over, gorgeous buttons. Fits me like a glove. Really flattering. The women who run the shop had noticed it too. So it had been given a special label to point out the quality to the casual shopper. Seeing that it was really special they had gone mad and marked it as an expensive item. It was £19. A high price for a charity shop shirt. But the way I see it, I didn't even pay for it. I made a charity donation and they gave me a shirt. I can't remember ever buying a shirt in a regular shop where I actually enjoyed handing them a £20 note and telling them to keep the change.

We have a planet to save and we're being radical. Some brilliant women have been thinking about radical changes to the fashion industry and the way we clothe ourselves for many years. In the following chapter I'm going to take you to meet Safia Minney, the undisputed queen of the sustainable, ethical, slow-fashion revolution. I hope I ask her all the questions that you'd like to ask.

CLOTHES AND FASHION 2
When Buying New Clothes is a Good Idea

The awards section of Safia Minney's Wikipedia page is really quite startlingly brilliant. Way back in 2009 she was awarded an MBE in the Queen's Birthday Honours list, and among her 15 other awards, there is hardly an award for promoting sustainability in the fashion industry that Safia hasn't been given.

One sunny morning I took the riverboat down the Thames to Tate Modern, to meet her. Safia is warm and friendly, with a huge smile and surprisingly petite at just 5 feet 3 inches.

I enjoyed our conversation so much that rather than counting it as 'research' and making it go flat with retelling, I thought I'd share the conversation exactly as it took place.

You can picture us. I'm a little nervous. I had wondered what on earth you wear to meet someone who has spent close to 30 years in the fashion industry. Deciding that there was nothing in my wardrobe that felt special enough to meet this accomplished businesswoman, I have chosen a black V-neck T-shirt (old, from Gap) and a pair of black trousers made from sustainable fabric (new). I'm also wearing bright turquoise heels to add some colour and energy. Safia (who is obviously not in the least nervous – why would she be?) is wearing a colourful cotton dress with small flowers, which looks relaxed and gorgeous on her, and some flat sneakers.

Safia, sensibly, just orders a coffee. I order food and then have to embarrass myself by trying to interview and eat at the same time. But she's so relaxed that I'm put instantly at ease. I have a little list of questions prepared so I attempt to put my professional face on. We are sitting on the eighth floor of Tate Modern, overlooking the River Thames and St Paul's Cathedral. Safia had arrived early to secure us a table with the best view.

I switch on my recorder, open my notebook and speak first:

'How should I describe you, Safia? Do I say, "Safia Minney is one of the first pioneers of sustainable fashion" or "slow fashion"?'

'I was one of the first people to use the term "slow fashion". That was through using a model of fair trade to deliver the best possible incomes to organic cotton farmers or artisans making textiles – hand-weavers, hand-knitters or hand-embroiderers in the so-called "developing world" – to make clothing that was the most socially impactful, had the most value added in it, and was the least environmentally impactful.'

'So you founded a company – People Tree – to make clothes in this way?'

'I did.'

'And how long ago was that?'

'Twenty-eight years ago, in Japan and the UK.'

'And that was very successful. And I think it's fair to say that your initiatives impacted the fashion industry as a whole?'

'Yes. I think so. We developed Fairtrade standards. We worked with global organic textile standards to develop what organic and Fairtrade cotton should look like. We created campaigns raising awareness of how exploitative garment manufacture was. So I can say that I was one of the first generation of ethical consumers that were horrified by sweatshop production of sneakers, denim and clothing. The idea that fashion and clothing had to be done differently was around pre-me, but in terms of making a commercially viable offer that was 100 per cent Fairtrade, that was using craft skills and carbon-neutral methods – People Tree was definitely the first in that respect.

'We did massive amounts of campaigning: we would run fashion events showing that carbon emissions were an issue, holding up placards talking about the need to cut CO_2 from production; that child labour was an issue prevalent in conventional fashion. We promoted Fairtrade principles, calling for gender equality, fair wages, transparency and accountability, and that growing organic meant multi-rotational crops for farmers. We'd be doing these fashion shows next to the World Trade Organization or the G8 in Hong Kong and at big green festivals in Tokyo or London. We made TV documentaries and films to report back from the slums to the consumers to tell them how garment workers were living because of the incredibly cheap prices they were paying in conventional fashion outlets.'

'To join the dots in our minds?'

'Exactly. We'd take documentary makers and influencers to the villages to show them the difference that Fairtrade can make and give them a very clear understanding of the industry.'

'Brilliant.'

'I'm an optimist and I believe that there is no point in moaning about stuff. You've got to deliver a viable alternative.'

'Yes, absolutely. I did watch the documentary *The True Cost* and I'll be recommending that all my friends watch it. Even though it's a few years old it couldn't be more relevant.'

'What conclusions are you reaching about clothes, Isabel?'

'The model that I've come up with, that works for me (bearing in mind that I hardly buy anything at all) is that I'd like my clothes to be at least 50 per cent second-hand and 25 per cent sustainable, "slow" Fairtrade brands. For the other 25 per cent, what I'd love is some clothes made either by me or by someone that I love. I've noticed that wearing clothes when I know the people personally that have made them, feels quite different. What would be your proportions on these choices for yourself?'

'I wouldn't buy anything new at all unless you really, really need it. And then, of course, if you do buy – buy sustainable and ethical from the pioneers. So you buy organic Fairtrade trainers from Po-Zu and you buy an organic cotton dress if you really need one. But if you wanted something special for a best friend's wedding or a once-in-a-lifetime event, you wouldn't buy it at all. Instead, you'd rent it from Wear The Walk, if you're in the UK. Or Rent the Runway is good in the USA.'

'I haven't considered rental at all.'

'I have a subscription with Wear The Walk so if I have a gala or some important event to go to I might rent an ethical piece for that. It's crazy for anyone to spend a fortune on an item of clothing that they may only wear on one or two occasions. And a lot of people are setting up clothes libraries – I know there is a Library of Things in Oxford – and there are clothes parties and just sharing clothes more with friends.'

'A Library of Things?[19] How logical.'

'Yes, people can borrow power tools or, say, children's toys. Because many children only want a certain toy for a couple of months and then they get bored with it.'

'And there is another advantage. I remember a Tibetan monk I met in Nepal once said that, in an orphanage he oversees, the children's toys are not individually owned because too much "mine" leads to too much "me".'

'They share everything?' Asked Safia?

'They do. So, a Library of Things sounds excellent for that too. Then there are "swishing" parties.'

'Swishing?' I asked.

'It's clothes swapping, including bags and anything else you like. It's a whole movement. Look it up online.'

'That makes good sense because if we are buying less but better quality then there will be fewer cheap clothes and better-quality clothes going to the overstocked charity shops. We know that, at present, charity shops have to get rid of the majority of donations. I'd rather give good-quality clothes that I don't wear to my friends.'

'Yes and these cheap clothes are often shipped, in huge crates, across the world to the so-called "developing world" where they completely destroy the local clothing market.'

'Another reason for us to pass on our better clothes to our friends instead.'

It's also fun having parties like this. Spending time with our friends, dressing up and saving resources all at the same time.'

'That's brilliant and I'm also keen to spend time with friends, learning how to make and repair clothes. The reason I've included this element which ideally would be made by me is that I realize that, like most of my friends, I've never made an item of clothing in my life – not for myself or for anyone else. I've mended clothing but that's not the same. I think it's important to understand how our clothing is made because it helps us appreciate it more.'

'That's a great idea.'

'How did you become interested in ethical clothing?' I asked.

'I think that, as an ethical consumer, every pound you spend should be spent on sustainable alternatives. If you want to be part of the solution you shouldn't be putting your money with the 500 corporations that are causing such huge problems.'

'I've been buying clothes for so many years only half aware of these issues and just wondered guiltily who made my clothes and in what country. I have wondered if my clothes were made in sweatshops and it has taken the joy out of clothing for me. I do want to know who made my clothes. Wearing clothes could and should be an expression of our personalities and of joy.'

'Exactly. My grandmother was an embroidery designer and so I had a background of beautiful Swiss linens with intricate embroidery.'

'So that's where your interest in detail comes from?'

'Also having spent 20 years in Japan. There they all enjoy attention to detail and they have a heritage of appreciation of textiles. Japan has given me a very different view of textiles because when we launched People Tree in Japan our customers would really appreciate hand-woven textiles. They'd turn a garment inside out to see the stitching and they'd really want to understand what kind of thread count it was and what kind of dyeing technique had been used. At this same time, Britain had forgotten anything to do with textile quality because everything was synthetic. In Japan they had a real celebration of the artisan as national heritage and treasure. They still have this in Japan but we don't in the UK. Enjoying textiles is all part of bringing back the love and the joy in wearing clothes.'

'I love the sensuality in what you're saying. It makes me want to choose clothes differently for so many reasons.'

'Exactly.'

'Do you have favourite ethical brands? Do you have any that you particularly recommend?'

'The more artisanal ones? That's a good question.'

'One problem is that if someone is new to all this and would like to switch to more sustainable and ethical clothing, there is so much "greenwashing" going on – brands pretending to be green or talking about the one element of what they are doing that is sustainable while ignoring the larger picture … it can be confusing.'

'I'd just go onto Common Objective online. All the pioneering ethical brands are on there. If I were a buyer at Topshop and wanted to source a factory that makes organic cotton I could find information there as a first stop.'

'Interesting that you mention organic cotton, because some people keep telling me how bad organic cotton is.'

'I think that's because it's largely confused with cotton.'

'It's an easy mistake – along with "organic food", formerly known as '"food".'

'Yes, but conventional non-organic cotton production uses a lot of water. Organic cotton uses much less water but of course it still uses water. Linen, which is based in Europe, is very sustainable.'

'These trousers I'm wearing are made of Tencel. Apparently, it's good on the ethical scale as it's made from sustainable tree pulp. I bought these from a small independent shop that specializes in slow fashion. Tencel feels fantastic. But are there disadvantages with it?'

'Yes, there are. It's got lots of lovely drape and we're very used to the way synthetics drape. So, Tencel is really great from that point of view. But the disadvantage is that it's a machine-produced fibre and I think we also need to think about how we create livelihoods for the poorest of the poor on our planet. If you look at the GDP [gross domestic product] of most West African countries, they are all cotton-based.'

'So you think that organic cotton is better than Tencel?

'I do. Yes.'

'Presumably, if you look at the environmental impact, then Tensel is less environmentally damaging than cotton, isn't it?'

'Yes, if you just look at the tree pulp and you as a First-World person wearing it. But if you look at the bigger picture you have to consider how to support 7.7 billion people on the planet. It's the First World that has created the problem of climate change so you have to consider what climate justice would look like. You have to look at how we create livelihoods for the poorest of the poor. Organic cotton would also be sequestering about 1.5 tons of CO_2 per acre per year in the soil. A hand loom would be saving about 1 ton of CO_2 per year, per machine.'

'I'm so impressed that you have these figures on the tip of your tongue. I understand what you are saying: climate justice is also about human rights.'

'It's about resources use and how we deliver basic human rights within the confines of the natural environment. So at People Tree we've aimed first at highly desirable clothes. Japan will never buy because of the sympathy vote – because it's Fairtrade – they will only buy if it's good design and good quality. But we also want it to have the maximum positive social impact with the minimum negative environmental impact. We do the global social and environmental maths. So that's the long answer to the question. Tencel is a perfectly good fabric environmentally but for social justice reasons I'd rather buy organic cotton.'

'So, in conclusion?'

'We need to encourage women to shop more like men. To have a capsule wardrobe and to think very carefully about what is in it.'

'Yes, to really love everything in it. So women no longer have that apparently ubiquitous complaint that they have nothing to wear.'

'And to bring back the fun in clothing, like in the eighties, clothing as a highly personalized way of expressing yourself. We seem to have forgotten that.'

'Yes. Everyone looks much the same as everyone else and it's easier just to wear black. Including me today. It's so boring. Tell me about your shoes.'

'Just as you can buy sustainable clothes you can buy sustainable shoes too. These are from the company Po-Zu. I was managing director there for two years after People Tree. They are made from organic cotton with Fairtrade rubber.'

'That's excellent. I've had a problem with shoes because I prefer not to buy leather, as I can never know how the animal that formerly owned the skin lived or died. But now I certainly don't want to buy plastic either.'

'Exactly. It's bollocks to say that shoes are OK if they are not animal skin. If they are plastic then they take 400 years to biodegrade. So Po-Zu would work well for you.'

'I literally can't wait. My trainers are all full of holes and falling apart because I've had them so long – but I haven't replaced them because I haven't found a source I'm happy with. Problem solved.'

'Brilliant.'

'What do you think about the Extinction Rebellion campaign to boycott fashion completely? Do you think it will influence consumers who are already thoughtful and so will damage small slow-fashion companies?'

'Their idea is to not buy? To boycott fashion completely?'

'Yes.'

'I know a lot of people who work in the fast-fashion industry who have also pledged not to buy any fashion this year. It's deeply ironic that their pay cheques are coming from a system that they are complicit in – and yet they themselves won't buy fast fashion. But I certainly don't think we should boycott ethically produced and sustainable new clothing because that is the way to deliver social justice. But I'm deeply supportive of any action that questions the mass production and mass disposal of clothing. In the same way that you are talking about the joy of treasuring something and wearing it forever. In the same way that Stockholm has stopped its Stockholm Fashion Week, it would be interesting to see something similar happening in London so that we can really question the whole machine.'

'That would be a relief.'

'If you're in the fashion industry you need to get up in the morning and look at everything through a climate lens …'

'I'd argue that whoever you are and whatever industry you are or aren't in – you need to look at everything through a climate lens.'

'This is true.'

We pause and drink our coffee. I refer to my last question.

'Safia, are you frustrated about how seemingly little progress has been made? I mean, you have been shouting about all this for 25 years and now, finally, hopefully, this is reaching a broader audience.'

'It's frustrating that, had a lot of the key reports not been covered up by vested interests and fossil fuel businesses, we might have had a good chance of limiting the damage of climate collapse and societal breakdown – so that's deeply saddening. But I think it's incredibly exciting that we are now creating the kind of society that we have always wanted to live in.'

'You say that but if we were meeting in Oxford Street and watching people going in and out of Primark, wouldn't we be asking if any impact is really being made?'

'You're right. It's not large enough at all and that's why movements like Extinction Rebellion are so important. That's why we can't wait to say, "Young kids know what they want. Let's wait for them to grow up and change the industry." That's bullshit. Middle-aged people have to wake up and realize how complicit they are. And, if necessary, redesign the way they think and the way they earn their money. It's absolutely key that the middle class and the professional class need to realize, without anyone blaming anyone, that they are part of the problem. And the working class too. Everyone.'

'Unless we live totally differently.'

'We all have a massive transition to make'

A massive – and I hope a joyful – transition.

THE LAW 1
Get Up Again, Over and Over

In Charles Dickens' novel *Oliver Twist*, published in 1893, the character of Mr Bumble declares famously that, 'The law is an ass.'

I had this idea, in my desire to give more tools for success to environmentalists, we could all learn how to use the law a little. But I very soon learnt that – where environmental law is concerned – to call it an ass would be over-complimentary since an ass is a useful animal that has been helping humans carry heavy loads and move both people and belongings from place to place for centuries. Environmental law on the other hand appears mostly to be less useful.

Emma Montlake, an environmental lawyer from the charity Environmental Law Foundation, ELF for short, explained it to me like this:

'You could have an incinerator that's being built. So what are the legal processes? You need an environmental permit. What does the permit permit you to do? It allows you to pollute the environment – up to a certain extent. If you go over that, and anyone notices and can prove it, perhaps enforcement will come in and you'll be prosecuted. But – on the whole – that's what environmental law does. It allows environmental destruction.'

She and I both stared at a beautiful beech tree rather glumly. I had travelled to Lewes on a hot October day for a delightful walk along the River Ouse. White cliffs dazzled in the sunshine offset by a clear blue sky. It was early autumn but very few of the leaves were yet turning gold. We were surrounded by deep green lushness. Her two gorgeous fluffy lurchers bounded along enthusiastically while I had an hour to learn about her charity and how they help environmental groups up and down the country.

'The truth is that cases are very hard won and, when won, usually need to be won again.'

Emma was a warm smiley woman, about the same age as I am – no pretence about her at all. Very honest, very immediate.

'I'm not sure I understand.' I was befuddled: 'If a case is won, why would it have to be won a second time?'

'For an example, there is the case of the Park Road Allotments in Isleworth, London. They are historic allotments that have been there for over a hundred years and are dearly loved, not only by those who maintain the allotments but by the entire community.'

'Oh dear – I can guess what you may be about to say next.'

We scrambled down a muddy bank.

'The land is owned by the Duke of Northumberland who also owns Syon House – one of his many properties. He decided that he wanted to build a hundred flats on the land for private rental. There wasn't any social housing in his plans. It was clear in his planning application that he wanted to build on the allotments. He offered the allotment owners a new space in the grounds of Syon House.'

'Even though many people have been tending their patches for years?'

'Yes. And there are many ancient fruit trees there too which shows the ground has a market garden history. He thought that a different bit of ground was much the same. The case went to a public inquiry and his appeal was refused. The reason for the refusal is a bit lateral. It was because he isn't allowed to dig up his own grade one listed garden to make new allotments.'

'So – happy allotment community?'

'Yes, but this is my point: four years later, the Isleworth Society comes back to us – another application has been made. A bit of social housing has been added this time, but again he is going to destroy their allotments. It goes through planning and again the planning application is refused. He appeals it. You can't imagine the cost and the amount of money the local community had to raise to defend their ancient allotments. We managed to secure them two fantastic barristers who did the whole case pro bono including the advocacy at the public inquiry. The inquiry lasts a week and that's a lot of everyone's time. Anyway the inspector decides that the Duke of Northumberland can't build on those allotments. So we won a second time. But will he come back a third time? He might …"

'I see. So this is why environmental campaigners have to win more than once to protect whatever bit of nature they love.'

'That's it.'

At that exact moment we passed a group of volunteers out planting chalk-loving plants. Celebrating this piece of nature. I hoped, for their sakes, no one was planning to build a car park here.

'The sad thing is that when the environment comes up against the needs of the economy, the environment rarely wins.'

'So tell me – are you an environmental lawyer or is ELF a charity?'

'We are a charity. Personally, I'm an environmental lawyer but what we do is we give advice. We don't litigate but we write letters. That's mostly what I do. Write letters.'

'That's funny. I guess many of us mainly look at computers a lot. Do you write real old-fashioned letters that you post?'

'No. Email only. They are called Pre Action Protocol letters, which is a legal letter reminding people of laws and statutes. A lot of what I do is force local councils to do what they are supposed to do.'

One of her lurchers threw itself into a large pond. Then clambered out and shook itself like a cartoon.

'What sort of letters do you write?'

'Would you like some coffee?' Emma produced a flask and two cups.

'Yes please. It's really beautiful sitting here under the trees in the sun.'

'A recent example is a case in Moylegrove, Pembrokeshire National Park. There has been a battle between those that want to use the area for recreation, kayaking and coasteering (which is basically swimming and clambering over rocks). So these people want to be in nature, which is fair enough, but they cause significant environmental damage. There is another group that want to protect the area for nature. In this particular area, Cardigan Bay, there are grey seals, little auks, a myriad of sea birds that nest there; it's a European protected site.'

'Please don't tell me that someone wants to build a multi-storey hotel on Cardigan Bay?'

'Not quite. But there was a planning application to renovate a bus depot to become an 'activity hub' for yet more recreational activities. The locals have already seen this increasing over the last ten years. They have gathered evidence for the local authority of the damage being done. I'm afraid that authorities around the country often put their fingers in their ears and sing "La La La".'

'And so how do you stop this?'

'In this case the activity company had just applied to convert the bus station and the local council were considering the application without considering the impact on the wider environment. So we got involved and wrote one of our letters pointing out that because of the protected nature of the site they had to have a habitats regulations assessment.'

'Good.'

'So Natural Resources Wales look at the plan and admit that we are right. So they do one and it's completely rubbish.'

'Can they get away with that?'

'Not on this occasion. We went back and said, "That was completely rubbish." We had some help from one of the lawyers who helps us. And so they had to do the environmental assessment again.'

'Fantastic.'

'And then two days later they passed the planning application.'

'Oh. That's not the ending to this story I was hoping for. No protection for the seals then?'

'I know. But because there was so much media attention to this, Wild Justice are now involved and are exploring a judicial review.'

'Wild Justice do great things, don't they?'

'Well, Wild Justice and the Northern Ireland Badger Group have managed to stop badger culling in Northern Ireland as a result of legal action.'

'So your action led to Wild Justice getting involved?'

'That's right. What we did was just force the local council to do what they were meant to do. We are just a small charity but we can do that. Local authorities have lost so much of their expertise that we hold local authorities and the regulators to account. We have to say "Hello? This is what you are meant to do and you're not doing it."'

'Do you work a lot with local councils?'

'We are a pro bono environmental law charity so what we do we do for nothing. But we have a network of barristers, lawyers and technical consultants who join our network in order to give more pro bono options to those who come to us.'

'So where does the Environment Agency come into this? I've heard they are very weak. Superb coffee by the way.'

'Thank you. The EA was completely defunded by the Conservatives. They don't have much money. It's clear, for example, that water companies have been illegally discharging sewage for decades and there has been very little enforcement by the EA. It's only now that there have been so many campaigns and citizen scientists all around the country gathering indisputable evidence that very hefty fines have been given. One of the things we sometimes do is help a small environmental group draw up a letter to the EA.'

'I thought it was the Environmental Protection Agency's job to – er – be the agency that protects the environment. The clue is in the name, isn't it?'

'EPA is the American version. Here they are just called the Environment Agency. It is only the Environment Agency that can bring enforcement action in England; in Wales it is Natural Resources Wales.'

'By giving out fines?'

'Corporations don't like losing money.'

'So another job for environmentalists now is to lobby the government to re-fund the Environment Agency so that they can prosecute the polluters?'

'In the spare time we don't have. Yes.'

We reached a bench and decided to sit in the sunshine. The sun was warm on our skin.

'So what else do you do besides writing to the Environment Agency?'

'I don't want to give the impression that we just do one thing. We have over 300 enquiries for help a year and the requests are very varied.'

'OK, so tell me about another one in which you've been successful. I like success stories.'

'One case where we were able to help was Manchester Ship Canal vs United Utilities. They are one of the water companies that have been very much in the news for all their illegal discharging activities.'

'Yes. Polluting Lake Windermere has been one of their specialities, I believe.'

'That too. And this case – from the Manchester Ship Canal Limited Company – who were fed up with their canal being polluted over a long period; the litigation has been going on for 14 years.'

'This is not encouraging. Why so long?'

'It's been up to the Supreme Court, back down again, and up to the Supreme Court again, and back down again.'

'Meanwhile the pollution goes on, all that time going into the canal?'

'Yes. In this case ELF was an intervener, which means that we were not party to the proceedings but you can intervene in a case if you think you have something interesting to say that wouldn't otherwise be brought up by the parties.

'It was a very technical case and the Manchester Ship Canal Company wanted the wider impact of sewage discharge on the surrounding communities to be brought into consideration. As ELF work with a lot of local communities, this is our area of expertise. So we had something to say about the very serious recreational, environmental and public health impacts of the sewage discharges. We applied to the court to intervene and they gave us permission to do so.'

'If it's not a stupid question – why are water companies allowed to dump raw sewage in public lakes and rivers anyway?'

'It started because under the Water Industry Act of 1991 it was held that impunity was given to the water companies to discharge sewage and there could be no legal recourse under either private nuisance or trespass for those people that suffer because of the sewage pollution.'

'Can the law be updated if it's no longer fit for purpose?'

'Exactly – there was a case to be made that part of the law was no longer good.'

'So what happened?'

'The Supreme Court said that there was nothing in this act that should give the water companies immunity. So finally this year there is case law from the Supreme Court that says if you have property rights over a stretch of a river (which are called riparian rights) and you suffer illegal sewage discharge into your bit of river then you can bring a nuisance act. Does that sound really boring?'

'No – not to me – it sounds like a huge success. Now we just need many members of the public who have property rights over a river to start bringing claims against the water companies. So what happens now?'

'We are waiting for someone with riparian rights to come to us saying that a polluter is causing a nuisance to their bit of river. But it can't be tidal. This part of the Ouse, for example, is tidal. It has to be a non-tidal stretch. No one has made a claim yet.'

'I'd be on that immediately if I owned a bit of river. But how would they know all this? This is why I think that learning the basics about how the law can help environmentalists is so important. I'm really interested in understanding how we can help with the implementation and enforcement of the law as well as the need for new and better laws. Someone may be despairing about the state of the pollution in the bit of river that they may be fortunate enough to own at the bottom of their garden and not even know that this new opportunity to make a case against the polluters exists.'

'That's true, and remember it's not just the water companies. Increasingly it's chicken, sheep and pig factories who are polluting the rivers with run-off, and pharmaceutical companies, and there is a huge dairy company in Cornwall – that sounds so innocent as they are just making cheese – but the environmental damage is horrendous and ... I could go on.'

I glanced at my watch. Emma and her dogs were being generous with their time. But it was a beautiful day, the dogs were happy and the sun was shining. Emma didn't appear to be in any great hurry to return to her letter writing and the 300 and more demands on her energy.

'So if people have a problem where they suspect that the law is being broken, they can get in touch with you and you can advise for free?'

'Well, we welcome donations from communities – but yes.'

'If I wrote to you about the poor air quality outside my house?'

'We'd write back and say, "There are statuary laws for nuisance which mean that if the cause is a local factory you can demand that the local authority investigate." They would need to come to your house and measure the air quality – they would need to do a complete investigation. Then if the local factory is breaking the law, they may be fined and after that it's all about enforcement again.'

'So how can individuals help with enforcement?'

'By gathering evidence.'

'So if you suspect that a local incinerator is ignoring the law, you have to collect evidence yourself in order to go to a law firm and say, "Here is the evidence – please take these people to court"?'

'Well, you'd go to the Environment Agency first.'

'And if someone has decided to build a horrendous pig-killing factory next to a nature reserve?'

'Once a decision is made your only recourse is judicial review. If a decision is made by a public body, the law is the only way to overturn it. Or statuary review if you are challenging the secretary of state's decision. But the good news is that if you bring a class action, that is one where a community is representing nature – or nightingales, for example – you are no longer going to lose your house due to adverse costs. Litigation is complex. ELF and lawyers that work with us could advise on whether to go down that path or not. I'd start by writing a letter to someone. As I said, that's mainly what I do.'

We stood up, roused the dogs and started to walk back under the trees along the banks of the Ouse, in the direction of Lewes Station. I wondered at the patience that she must have to find.

'So it's a bit like "Man who climb mountain start by taking one small step." Except it's "Woman who help to save planet start by writing a letter"?'

'Yes. That's it exactly.'

And of course there are other cases where local community groups have achieved extraordinary successes.

Can you imagine finding out that someone plans to drill for oil on a local favourite hill of yours where you enjoy a walk with your dog? Then you learn that your local council has approved the plan.

This is what happened to one Sarah Finch, who discovered that plans had already been approved for a site in the middle of Surrey, which would include five drilling cellars, four gas-to-power generators, a processing area, an area for storage and tanker loading, seven 1,300-barrel oil tanks and a 37-metre drilling rig. All this was going to add to the large-scale production of up to 3.3 million tonnes of crude oil for sale and use as a fuel for transport for 20 years. And it was approved.

I would probably just have sat and cried at that point. Many of us would have concluded that if it was already approved by the local council then it was hopeless. But Sarah is made of more valiant material. Being a wise human, she knew that we can't do much alone so she started a group of sane people and they called themselves the Weald Action Group and they started reading. They needed to challenge that decision.

As you now know, when someone wants to make money by damaging the environment, before a local council can grant an application they have to make an environmental assessment report. They found out that, in considering the likely environmental impact of the project, they had only considered the damage that would be done by drilling the oil. If you can believe it, they had not been obliged to include the impact of the 'end use' – i.e. burning the oil.

Whether they could challenge this decision or not turned on the interpretation of the rules about impact assessments. As we remember, the fate of the ecosystems hangs in subclauses. They decided to risk it – they raised a lot of money using every way they could think of, including one woman in her 70s undertaking a 100-mile sponsored walk. They wanted to take Surrey Council to court. The courts are often not kind to environmental action groups. Her case was refused permission on the papers and again at an oral permission hearing. Again I'm sure I'd have given up at that stage. But when one of the group wanted to quit, another would encourage them to continue. This process took years. The case was granted permission in the Court of Appeal. Then they went to the High Court and lost again. They went to the High Court of Appeal. They lost again. Can you imagine the fund raising for all this?

To cut a long story short, they lost. Lawyers for Surrey County Council called Sarah and the Weald Action Group, 'misguided'. But they also had lawyers. The climate lawyers acting for Sarah – who I will name because I can: Leigh Day – didn't agree that they were misguided. Carol Day, their lawyer, tells me that she told Sarah she was turning the *Titanic* around. They decided, against all the odds, to go to the Supreme Court. They had been making the same point all along – surely the 'downstream' greenhouse gas emissions (those that will occur at a later date) logically have to be measured as part of the environmental impact.

Despite all the previous failures, the Supreme Court agreed with them. And they won. This case set a new legal precedent that has profound impact not only on the UK's ability to meet our carbon reduction targets but also, without exaggeration, on the planet. After this win, because a new legal precedent has now been set, it applies to any and all new projects that might involve burning oil, coal or gas. And not just new ones. This new precedent has given campaigners a tool with which to challenge existing coal and gas projects that are in development.

And how did Sarah feel after all this? In an interview she said, 'absolutely vindicated but also a little bit angry, because it's taken five years and haven't we said the same thing all along?'

So these kind of super successes don't happen often. But we can all help with the implementation of the law and, by our efforts, help enforcement to happen. Carol Day at Leigh Day, the Senior Environmental Solicitor at the company, is also very encouraging. She says:

'I would say that concerned individuals and community groups are the legal eyes and ears when it comes to enforcement. Nowhere is that clearer than in the sewage scandal. It has been groups like Windrush Against Sewage Pollution (WASP) and all the wild swimmers: these groups have managed to use one of our lesser-known, but very effective, environmental laws – the Environmental Information Regulations 2004 – to obtain data and information that would otherwise not have been in the public domain. These groups have become extremely knowledgeable about the issue and proficient at using these laws to achieve their campaign objectives.'

So maybe I was a little hasty. Maybe the law isn't always an ass. Maybe it's me who is the ass because, unlike all the members of all these groups, I've never learned how to make use of the law effectively and joyfully to become a better environmentalist. Moral of this section – don't be like Isabel. If you want to become a more confident and more competent environmentalist – learn about the law.

ENERGY 7

The Environmentally Friendly Hearth

I live in a home without a fire. On a cold winter's night, in order to maintain my joyful environmentalism I have been known to enter the words, 'Relaxing fireplace with crackling sounds' into Ecosia. I place the open computer down in the hearth, where a fire isn't, before settling down in front of a video of cheerful flames. Visiting friends often love this – it sounds utterly convincing and is strangely deceptive to the senses; it's comforting and relaxing. The problem is that it offers no heat. My daughter sighs wearily when I put on a virtual fire and calls it 'tragic'. Ha ha ha.

I amuse myself in this way, but who would choose not to have a real fire? I would never again choose a property where on-line flames are the only available option. But having a fire and being an environmentalist can be confusing. The government has brought in new laws and made various statements about pollution and what can and can't be burned and how it must or must not be burned – creating more befuddlement than clarity. So, can we be environmentalists and still enjoy real fires?

I got in touch with the UK-based Stoves Online, one of the companies that lead in this field. They successfully unmuddled me. So here is the low-down for you.

Open fires give 20 per cent of the heat produced to the room and the other 80 per cent goes up the chimney and into the atmosphere. This is not what you want for your heating bills, your home or for the planet. But if you have a wood burner of the correct design then 80 per cent of the heat is delivered to the room. From 2021 there was a new law in the UK that you had to install a registered and approved 'eco-design ready' stove.

The new laws prevent open fireplaces and old stoves being put into new properties because they are not energy friendly. These two methods of burning wood, combined, account for 50 per cent of the wood production in the UK.

Electric heat pumps, as Juliet Davenport suggests at Good Energy (see page 260), are great if you have a supplier that uses 100 per cent renewable electricity and you can afford them. But you're not likely to have a good glass of organic wine on a cold evening and snuggle up around a heat pump. There is a romance with fires that makes a house into a home – but if you're going to keep your fire and be a good environmentalist you need to make sure you are meeting the 'eco-design ready' standards.

Don't burn coal. Ever. It's acrid and apparently cancerous fumes come from coal. And don't burn wet wood because that's what produces more of the smoke that contains 'PM2.5 particles' which are bad for you if you breathe them in. Make sure your wood is fully dry before burning it.

As to the wood you are putting into the fire, like everything that comes into our homes we need to look at the supply chain. Importing wood by the crate load from abroad is inexcusable. Use locally sourced coppiced wood, which has been collected for centuries, that's supporting small dealers who care about the management of your woods as much as you do. Once again, as with the supply chains of everything we use, we need to know exactly where the wood is coming from.

If you live in a city in the UK, you have to use a Department for Environment, Food and Rural Affairs (DEFRA) approved stove for smokeless zones. This is another test that the stoves have to pass so that they can be used in smokeless areas — which is pretty much every city in the UK.

From the environmental point of view old cast-iron stoves that we stroke and speak of lovingly are a problem. Stoves Online offer a 'scrap-in' scheme to prevent people 'up-cycling' them with a coat of paint and reselling them to innocent punters who don't realize that they don't meet the new environmental standards. All other UK dealers are expected to follow and offer the same 'scrap-in' scheme.

Taking a deep breath I asked Stoves Online, 'And how much money does a brand-new eco-design ready DEFRA-approved stove cost?' Apparently, you can spend between £700 and any amount of money you'd like to spend according to the design you want to look at.

So, my best research tells me that if I ever move, I'd heat my house with an electric pump with energy supplied by a company with 100 per cent renewable energy and to stop myself sitting in front of computers on cold nights I'd have a brand-new £700 eco-design ready DEFRA-approved wood burning stove. And I'd be very careful where the wood I fed it with came from.

ENERGY 8
If I Were a Rich Man: Heat Pumps & All That Jazz

Fiddler on the Roof has always been one of the favourites in this house. Everyone here can do a passable rendering of 'If I were a Rich Man' from the classic film, including an unnecessary number of the words in the verse about wanting one staircase that would just go up and another that would just come down.

But these aren't my personal desires. I don't want a 'big tall house with rooms by the dozen, right in the middle of town'. I'm quite happy with the flat I live in. My dreams are a little more modest. So what would I want, the curious reader might ask? Ready to sing?

'I'd just buy a heat pump. Ya ba diddle diddle diddle diddle diddle diddle dum.'

The song would go on: 'I'd have a friendly little white pump whirring away' (I could sing this version) and 'lots of solar panels on my roof'. Those of you that know the

IF I WERE A RICH MAN: HEAT PUMPS & ALL THAT JAZZ

tune will be humming along by now. But alas, my dreams, like Topol's, are unlikely to come true. For me, heat pumps, like swimming pools, are something that I see only in other people's houses. I visited a man near here with one on his conservatory roof. And I know of an environmentalist locally, who, despite being in her 90s, has just had a heat pump put into her house.

'I won't be here to reap the cost back,' she said, 'but whoever lives in the house after I'm gone will have a very low carbon footprint.'

Perhaps this is the 21st-century equivalent of planting trees so that other people can enjoy the shade. There are some seriously good people out there in the world. I often think about this.

What amazes me, on the other hand, is knowing that there are people that care about the planet and have enough spare money to retrofit their homes – but they just don't get around to it. I mean – we have climate change happening. This is an emergency. Didn't they get the memo?

If we want to cut our personal impact on carbon emissions by 65 per cent or more – the most important step we can make is to stop heating our homes with gas or oil. I'm just going to write that again – 65 per cent. And what's frustrating is that some people can afford this. They could fit double glazing, solar panels, and have a heat pump to heat their cozy home and still be able to afford a bottle of bio-dynamic red wine to drink responsibly at the end of the week.

I know this because some people buy cars. Some people spend mind-bogglingly large amounts of money on cars. Some people buy yachts. Some people buy second homes and still haven't bought heat pumps or solar panels. This isn't even logical because people who have money are usually very financially savvy. And buying a heat pump with some solar panels will pay for itself in about eight years. Not only is this the right choice for the planet – it's also the right choice for our personal bills. You just have to do maths for the long term.

And the thing is, friends – it's free to research the options. It doesn't cost anything to find out what grants are available in your area. Mostly, at the time of going to print for the second edition of this book, grants of £7,500 are available and they are likely to go up to £10,000 next year. The UK government is keen to save money and meet our emissions targets. They are retrofitting social housing every day.

This morning a very charming man called Sam came to my house and told me that his company are currently fitting about ten heat pumps a week. Of those, five or six are into social housing and the rest are into private housing. Anyway, he walked around, both inside and outside, and told me that I could indeed have a heat pump. This was happy news as several energy suppliers had told me, 'We don't fit heat pumps in flats.'

His company also told me that you don't have to pay all at once. They said, 'We don't work with a financing company but you can pay in four or more chunks. We look at each person's circumstances.'

I asked Sam all sorts of questions.

'So how does this work then?'

'Very simply, it's like a refrigerator or an air-conditioning system in reverse. It takes air from outside your house and heats it up.'

'Could they fit the pipes without having to take up the floors? (Yes, they could). Could they put solar panels on the roof and then just run a cable down to my flat because I don't have the top flat? (Yes, they could.) Would it be better if I found a way to fund double glazing first? (Yes, it would.)'

'It's all about heat loss,' said Sam. 'You have to look at your heat loss first.' As you know, I've looked at this before. But there is more to do. I'm still losing heat through the cat flap, installed so the neighbour's cats can come in and out when they choose.

And so I'm recommending a project for me and for you. Get the heat pump plan costed. If you're interested in saving our planet – and we know you are or you wouldn't be here – just do the costings on your home. If you're in social housing find out what's happening in your part of the country. Some energy companies want to charge you for the assessment of your property and then take the cost off if you proceed with the fitting. Some companies will charge you for the assessment but give you 100 per cent of the money back if you decide not to proceed.

This is a wonderful research project and think of how much we could learn just by exploring this. (Great for older home schoolers?) This simple question – 'Should I get rid of my old gas boiler and buy a heat pump?' – could be a university module: it covers not only science and environmentalism but also engineering, chemistry, maths, social psychology, budgeting, finance, business, ethics and design. I would include in that a bunch of life skills about being wise and shrewd, as prices are likely to vary so much as to take our breath away.

And then you need to know how to avoid going to a company where the fitters may not be properly trained. One evening watching YouTube videos on the subject will fill you in on all the horror stories. Some poor customers have a system fitted that costs them a fortune, then doesn't work and when they turn around to complain and ask the fitters to put it right, the company doesn't exist anymore. No one wants to have that experience.

My research to date has told me that Heat Geek – who not only supply to customers but also run courses for installation professionals – seem to be the most trusted company, as well as running some very entertaining social media. Also, turning to my trusted Ethical Consumer Magazine once again, I hope they won't mind me telling you that the most ethical pumps on their list are Grant, Mastertherm and Vaillant heat pumps. The bottom of their lists features Samsung and Mitsubishi – so avoid those two if you're buying one or check that the company you choose is buying one of the most ethical ones for you.

And not all the companies are cowboys. Sam (not from Heat Geek) spent the first three-quarters of his visit to my house showing me where I could put a heat pump, explaining where they could lay the pipes, telling me that his company could also deal with the double glazing and what it would cost to put up scaffolding to fit the solar panels. Then just as he was leaving – it was as if he recanted. He knew he was supposed to be there to sell his company but he eventually said, 'The thing is, Isabel, your gas boiler isn't old. It isn't one of the worst ones. It doesn't make sense for you to do all this until your present boiler stops working. And that may not be for another five years. The technology is changing all the time. It gets better and better. The pumps get quieter. They may be a lot cheaper by then.'

I said, 'Oh.' And felt a bit sad.

He added, 'And it doesn't entirely make good environmental sense to throw away a boiler that's still working.'

I suppose he's right. I would still do it. If I had more money. Because, regardless of all other criteria, I want to cut my CO_2 by 65 per cent. I don't want to use gas. At all. Ever.

I tell you what I'd do in your position, said Sam, suddenly going all fatherly.

'What?'

'I'd start saving. I'd create a savings account somewhere and I'd start saving. Because sooner or later your gas boiler will stop working. And then when it does, you won't have to have a cheap gas one – that is, if the government are still selling them as there is talk of them phasing out new gas boilers completely – well, then, you'll be able to get what you want.'

'Have I understood you correctly? You are saying that you could do me a heat pump installation, but you're advising against?'

'In your case – for this property – at this time – yes. Anyway, relative to some of the huge detached houses I see, and the amount of outlay compared to the small amount of energy you must use to heat a terraced flat – you can't afford one.'

And so I sang Sam my little heat pump song. And I danced like Topol as he laughed and drank his tea and didn't seem to mind at all that I'd wasted his morning.

But I do want to do everything I can to get off gas and cut my CO_2. So I'm going to start saving. And when my gas boiler dies, I'll be ready, diddle diddle diddle diddle diddle diddle diddle dum.

NATURE 1
Love, Peace and a Nightingale

Many people work hard at finding inner peace. Those who learn to meditate describe being driven crazy by the persistent bombardment of their own thoughts. There is a striving to switch off the constant noise. This continual brain chatter, sometimes running on repeat loops, is often loaded with worry about society, heavy with sadness, busy with anxiety about how to survive, troubled by concern for other people's problems – or, increasingly it seems, attacking thoughts directed at the poor 'self' who is trying so hard to look after us.

I've done my own share of mediation retreats that have involved torturing my body and mind just to achieve stillness. On a Vipassana retreat once, after eight days of physical pain and mental gymnastics for ten hours a day, I was finally able to enjoy sitting still for two-hour stretches without agony or internal complaint. But this wasn't easy to achieve.

Some admit 'I can't sit still,' so to find some silence they take up sport or some hair-raising activity that requires their focused attention. If you're climbing mountains at unbreathable altitudes or hanging, by one arm, from a rock face, it's essential to concentrate on your next move in order to stay alive. You're not able to worry about the state of the international economy, the injustices of the class system or your parents' divorce party.

There are the standard dysfunctional ways of taking a break from our thoughts: using drugs or alcohol; the lazy ways: watching too much TV or Netflix; the compulsive ways: an exercise class six days a week; the money-earning ways: working way too much. This list could get lengthy.

Then there are the positive ways: a stretchy yoga or t'ai chi class; the musical ways: singing in a choir or creating musical harmony; the noisy ways (as you know, Japanese taiko drumming is my fix); the nutritious and loving ways: cooking sumptuous food for your friends; the ecological ways: gardening, or even, in some cases, service to others in any form. All these things can move us away from too much thinking.

There are also numerous so-called 'spiritual' routes offering greater peace. An amusing panoply of different positions on and for meditation offers opportunities to observe your thoughts and then, if you persist with your practice, you can enjoy the increasingly longer gaps between them. Mindfulness encourages us to be fully aware of the present moment. Advaita teaches you not to identify with thought at all. People pay hard-earned money for courses to learn these practices. Peace, it would appear, is something most people feel we have to work at. Something we need to learn – and to do.

But ... I know a place in the UK where you don't have to find peace. Peace finds you. I'm sure there are others across the world, but personally I've never experienced anywhere quite like this. This is a place where a brave couple have done something that many consider remarkable. They have taken 3,500 acres of former agricultural land – and left it alone. I have brought three friends with me and we are sitting, on a rug in the sunshine, on land belonging to the Knepp Castle Estate in West Sussex, southern England.

Charlie Burrell, who owns this land, inherited it. It's been in his family for over 200 years. It's a huge acreage and farming it was a massive responsibility. Charlie went to agricultural college so that when he took over the farm from his grandmother in 1987, he was ready to implement all the most modern farming methods. He and his wife, the author Isabella Tree, worked hard running a large successful farm rotating the land between arable use and keeping cows for dairy farming. They supported a staff of 11 and their families. However, despite the fact that they worked harder every year and yields went up, their profits went down. Their results were an extraordinary success but they were getting further into debt. It became clear that no matter how efficient and productive they might be, they would lose an impossible amount of money if they continued intensive farming. After 15 years of trying everything, their arable and dairy business was finished and they had to find an alternative.

Professing what they called 'an amateurish love for wildlife', the Burrells decided to turn over their land to a pioneering rewilding experiment. Rewilding land is extraordinarily courageous and profoundly simple at the same time,. It requires human beings to do the one thing they find the most difficult: nothing. You restore the land and revive nature by allowing nature to take the driving seat. Non-interference. You don't cut back anything, sow crops or plough fields or even scatter the good seed on the land. This, my friends, is considered so radical that the issue divides farmers and landowners across the UK and across the world. The Burrells prepared to be hated by every overworked and exhausted farmer in the country.

Farmers are used to being in charge of every last inch they own. During World War II, anyone who had earth was encouraged to 'dig for victory'. Any 'fallow' 6 inches was considered wasted and constituted an unpatriotic crime. The earth, like the people, had to be 'productive'. So, the idea of deliberately making land 'unproductive' is not only going against the grain but against everything most farmers were, and still are, taught at agricultural college. And being British (or French, Spanish, German – or complete your own list), people are resistant to changing their ideas. Even if they are being slowly but surely driven to bankruptcy and know that their chemical and toxic agricultural methods have killed the very earth upon which their livelihood depends.

So, seeking what Isabella Tree later described as 'a more fulfilling connection with nature in its awe-inspiring and unfettered complexity,' they decided to get out of nature's way. Funding bodies that could have helped the recovery bristled. What were their 'targets'? Which species were they aiming to protect? What would they do to support these particular species? What were the predictions? The Burrells couldn't answer any of these questions. They were going to do nothing and see what nature wanted to do. That was the point of the exercise. Many of their funding applications were rejected. How could the funders say that targets had been met when they had no targets? No boxes to tick? This inability to measure outputs unsettles both scientists and funding bodies.

Many local neighbours went into meltdown. The Burrells' beautiful land, which had been so 'well maintained' by proud farmers for generations, was now being 'neglected'. They were a disgrace to the family name and, worse still, they were not even pulling up the weeds. Furious letters were written to local papers and funding organizations about this ruinous neglect. 'It's a visual disgrace,' letters raged. People had forgotten that intensive farming had previously altered the landscape beyond anything that their own grandparents would have recognized.

We suffer from what is called 'shifting baseline syndrome'. Our 'baseline' is what we expect our environment to be and to look like – but only measured by what it was like when we were children. Apparently, we are happy to accept a slight degradation from that. Who dares to point out that there was a time before our countryside was made up of fields filled with sheep? Everyone thinks sheep are native to the UK, as they have become associated with our childhood images. But decades before these fields were only grass, there would have been 10 different species of tree, 25 species of wild flowers, badgers, beavers, lynx, hedgehogs, otters, foxes, wild pigs, deer and more varieties of birds than you can count. This is challenging what is perceived as normal. 'Normal', in other words, is relative. We need to shift our baseline way further back than the pictures of sheep and cows from our childhood.

The Burrells patiently filed their hundreds of letters of complaint, did their best to reassure their neighbours in local meetings, made compromises and kept going. They took all the grazing animals off the land to allow whatever wanted to grow there to thrive. More letters of complaint arrived – now they were not even producing milk! What sort of deluded or ill-informed ideas had taken hold of them? Charlie and his staff had to put aside their egos, their agricultural training and their fattening folders of letters. They gathered their friends around them and allowed time for their land to become a truly natural habitat.

It took years for the previously intensively farmed land to begin to recover. But as the chemicals, pesticides and industrial farming methods became history, nature began to reward them.

One of the first areas of growth was what is known as 'thorny scrub'. Pretty much the entire list of all the trees and shrubs that you'd plant to encourage bird life started to grow everywhere. Most of it has thorns so farmers would typically have all this removed. Blackthorn, hawthorn, dog rose and bramble were prolific and, in the middle of these bushes, saplings of oak and alder grew up – protected by the thorny bushes around them. This is how trees grow naturally and this is why these bushes have thorns. Not to keep out burglars when planted by humans, but to keep grazing animals away from young trees until they are able to fend for themselves. Once you know this, you have to smile at people planting trees and using yet another product made by the oil industry to 'protect' them. If landowners would just let the scrub grow, then we wouldn't need plastic cones around every growing tree. How did nature manage before humans came along with plastic?

After years with no animals on the land the Burrells first reintroduced fallow deer, then Old English Longhorn cattle, then Exmoor Ponies. Later they brought in two Tamworth sows and eight piglets. These beautiful pigs are as close as possible to those that would have lived on the land before humans came and displaced them. Laws forbid the introduction of wild boar (even if the land is fenced), so they chose a breed of pig most closely related to boar and just let them roam freely on the land. Pigs, like dogs, love digging, which plays havoc with the land in ways that give conventional farmers nightmares. They dig up the earth, eat whatever they fancy and break up the river beds if they want a swim. They love swimming. Apparently they also dive for roots and rhizomes in the rivers and lakes. So, pigs, left to their own devices, know how to dive into a river, hold their breath, pull roots or even swan mussels from the bottom and surface again with their snack. A bit different from raising pigs in pens, isn't it? And the pigs benefit the land hugely. Did you know that the reason the garden robin is so friendly is because in its natural habitat it follows pigs around, knowing that they will turn over the earth and make it easier to find worms? If you've ever been in a garden and a robin has come up to you, it hasn't come to say 'hello' – it's hoping that you're going to be a good pig and start digging.

The deer, cows, ponies and pigs all hang out together. No sheep because – I'll just mention this a fourth time – sheep are not native to the UK. All the large herbivores continue to help the land to recover. But then, once again, this makes local farmers and other nervous locals panic. Surely the Burrells will be feeding the animals? (It takes a long time for people to understand that humans are not the gods that we thought we were.) Nope. But surely the animals would starve? Or die of cold in winter? Surely the animals can't actually look after themselves? To keep the panic-stricken locals happy, and to avoid being reported for neglect, the Burrells leave open the doors of a few barns so that the cattle can avail themselves of these human-made luxury facilities should they wish to. The animals say 'thanks – but no thanks'. They also stubbornly continue not to die of starvation or to freeze. On the contrary, these herds are not just eating grass, they are eating whatever they want to. This includes all the herbs, shrubs, blossom, berries, leaves and twigs or 'tree fodder' (from pollarded trees) that commonly used to be given to animals in winter in medieval times – and still is in some parts of the world. Animals absolutely love it. They know what to eat and what not to eat. They don't need humans to feed them.

If you're clever you'll have noticed there is a catch in all this. Laws require, for example, that calves have to be tagged within a few days of birth and if there is a national or regional outbreak of disease, DEFRA recommends vaccination. Knepp Castle Estate also treats them if they pick up liverfluke or have any complaint that needs antibiotics – but they take them out of the system, according to Soil Association Organic regulations, so that the soil remains organic. They then they put them back once the chemicals/medicine has passed through their system. And, sadly, that is not the only human interference that becomes necessary. There is the problem of breeding. (Miraculously, they all know how to breed and give birth without the help of vets.)

When I ask about this, Isabella Tree tells me, 'We have to castrate our bull calves and we've also castrated the Exmoor stallions in the past because, without a market to sell them, we couldn't let our herd grow too big. It was a painful exercise for everyone involved – not to mention the expense. We've now got an intact stallion running with the herd and, for the first time in years, four foals – which is lovely to see. We've been able to do this because we're going to increase the herd in the Southern Block.'

The hard reality of Knepp that I don't like is that they kill some of their animals. There are no longer any apex predators in Britain and, they explain, predators only account for a very low percentage of mortality in prey populations. It's food resources – the available fodder – and weather conditions (harsh winters, flooding, droughts, etc., which also affect the food resources) that are responsible for the boom and bust population curves of herbivores in nature. So, even if they are ever allowed wolves at Knepp, the wolves wouldn't keep the numbers low enough.

Without human killing, what would happen (as it does in some cases in Africa or in Yellowstone) is the numbers of herbivores would expand until they ate all the available food, then they would collapse, and as the grass and fodder recovered, the populations would begin to climb again.

It is claimed that when you kill animals in a system like Knepp, what you're doing is pre-empting that cycle of starvation and preventing them dying slowly and possibly painfully of old age and disease. I'm not so sure about this argument. I just hate to see a healthy animal killed for any reason. After all, you could kill a human and say that you were preventing them dying of disease or old age. But I go on listening to Isabella and trying to understand their system. At least these animals have good lives.

The Burrells do allow some of the animals to grow old because, for example, older matriarchs in a herd of cows apparently stop the young bulls from being aggressive. And heifers – the teenagers!

The matriarchs pass on knowledge down the generations: where to graze, what to look out for, the best places to shelter – so they want to keep the lead grannies. But how they decide which ones to kill and which ones to leave is a subject that, as a vegan, I was curious about. Apparently, they select which animals are consistently prone to things like mastitis and laminitis, and those that are neglectful of their calves or prone to liverfluke. They say that they want to ensure strong bloodlines that are also resilient and adapted to their particular location. Isabella explains, 'We cull in a pyramid system – so include animals of all ages – thereby preserving the varied generations of the herd.'

I can see they know what they are doing and why. But I still can't write the word 'cull' without feeling that humans are far overextending our rights as a species that not only should protect the voiceless ones but also needs to, for our own sanity.

And of course having killed the animals there should be two choices. In nature the dead animals should rot on the land, providing food for insects and carrion birds – but it's illegal in the UK to leave dead animals on your own land. The law requires that the tender humans be sheltered from the realities of life and nature – partly not to upset us and partly to do with the idea that rotting carcasses spread disease. But in a healthy ecosystem, on healthy soils, the carcass rots (and in the process sustains a whole army of necrophagous insects, birds, etc.) and returns very quickly to the soil.

It's only when carcasses rot close to water sources that they may create a problem. So, Knepp are in discussion with the Zoological Society of London about this. They believe that there's a strong case for allowing fallen stock to decompose naturally on the land, as long as it's away from water.

At the moment, the Burrells either have to break the law or take the other choice, which is to sell what some people call 'meat'. Unsurprisingly, they choose the latter.

I must admit my heart fell when I learned this. I found myself thinking, 'Oh, so you're just finding a different way to farm, kill animals and sell them?' And yes – that is exactly what they are doing. They would say, 'Yes we are farming. But way less intensively.' I'll admit that they have fewer animals per hectare than even conventional conservation grazing, and exponentially fewer animals than intensive farming. But as a vegan this doesn't sit well with me as I prefer not to eat animals or any sentient creatures, at any time. But I can see here that if they are going to keep animals on the land – and nature would have animals on the land – their choice is either to have ever-expanding herds, which would destroy the project completely as the land has to be fenced so the animals would starve, or to kill them. They could castrate all the males and leave only adult animals on the land but that wouldn't be natural either. I don't like this any more than any other animal lover. But I do see that they have what I consider to be the ONLY argument for eating meat that I have ever heard that I think has any legitimacy. And of course they are not farming intensively so all the usual arguments about the ecological destruction brought about by intensive animal agriculture don't apply here.

The Burrells say that they are very happy eating their own meat. But Isabella (a former vegetarian) says that she's happy because she knows the life that the animals have had and the part that they have played in the ecological restoration.

She does, however, admit to struggling with their killing at the local abattoir. It would be better for the animals if they could kill them at Knepp. So they are looking into the possibility of using a mobile abattoir on site. Apparently, that's better.

They do not sell dairy products as they don't milk the cows. Their daughter doesn't drink dairy milk at all and the rest of the family source organic, pasture-fed, unpasteurized milk – so it's not from a grain-fed, antibiotic-ridden, industrialized system. Although they will never milk their own Longhorn cattle, they do support small outlets that sell 'calf-friendly milk' where you leave the calf feeding and take just a percentage of the milk. Isabella has a section in her book *Wilding* that describes why dairy is so often a cruel industry.

This is the only aspect of the Knepp Wilding experiment that I'm not happy with. But then I'm not happy with my neighbour's cat killing the birds in the garden either. I guess I need to toughen up a bit.

Putting grazing animals on the land prevents it from turning into closed canopy woodland, that much I know. And with animals now back where they should be, the natural recovery of the land is further enhanced. The dung from industrially farmed cows is full of antibiotics and pesticides so that even when a nice big cow pat lands on the grass, dung beetles and other insects in most parts of the country would be well advised to keep away. They're killed by dung that is laden with avermectins and antibiotics. Not so at Knepp. The dung is a natural fertilizer for the land and almost as soon as one is donated, beetles fly in from who knows where, burrow their way in and start to pull it down into the earth to nourish the soil. The same happens with the various gifts from deer and horses. So the land is being naturally nourished, as nature intended, all day and every day. Insects start to party. And where there are insects there are birds.

The butterflies and birds have particularly appreciated this experiment in mankind fucking off (sorry to use this language but really that's mostly what we need to do). The birds are particularly grateful to have a piece of real nature to call home. As the land became itself again, naturalists and students writing theses of various kinds began to show up and find ways to count and measure the successes. On 2 July 2018 they counted 388 Purple Emperor butterflies on a single day. They are very rare in other parts of the UK. These were followed by an influx of butterfly enthusiasts. And with the arrival of ever-increasing numbers of birds came the opportunity to offer visits to birdwatchers. As nature recovered more and more each year, the Burrells had people from all over the UK and all over the world who wanted to come and study, watch birds and simply experience nature the way that nature can be.

I have never been a birdwatcher. I know the names of most of the birds in my garden, but I can only identify two or three by their song. But when I found out that Knepp offers nighttime walks to hear the song of a nightingale I found myself buying four tickets for an evening six months in the future. But six months passes quickly. And here we are in the sunshine. Sitting on grass that's part of the less than 1 per cent of land that is preserved for nature in the UK. In a little corner of the Knepp Estate that is now a camp site. The Burrells make sure that there are never too many visitors at once so we feel as if it's all ours. And in this beautiful present moment, it is.

Clare is a young farmer from Derbyshire. Her family has worked the land traditionally for generations. It's almost mean of me to bring her as I know that the implications of this trip will separate her from the thinking of all those around her – her family and friends and all the young farmers that she knows in her area. I bring her here for a wonderful experience and then leave her to try to work out what to do with the implications of all this knowledge.

My second guest is my daughter, usually the life and soul of any gathering. She's exhausted from her crazy London life and I've really brought her just so that she can drink in the peace.

And I've also invited Zoë, my friend from Extinction Rebellion. Zoë – like many environmental activists – is suffering from a kind of mild but chronic climate anxiety. Many people she knows are becoming desperate to help the environment because they see nothing is changing for the better. I want her to experience that, here at least, nature is thriving. We are all here to listen to the nightingales sing.

What's interesting though is observing what happens to us. We arrive, we pitch our tent and then we just sit on the grass. Peace comes and grabs us all without any effort necessary. We don't even have to cross our legs and think about breathing. We just sit on the rug on the grass and I watch peace take hold of each of us. I feel it in myself. At once the most profound and the most simple of experiences.

It isn't just the beauty all around us – although all that green is certainly good for the soul – it's the birdsong. A kind of birdsong that none of us has ever heard before: 100 per cent birdsong.

There is a cliché that is often thrown at environmentalists: that we want to 'take people back to the dark ages'. But perhaps the dark ages were not dark at all? Take the Asháninka native tribal people in the Peruvian rainforest: their way of life may be 'unsophisticated' in that it lacks technology, but it is happier than ours. So I ask myself, as we sit in the sun, whether this remarkable light, sound and peace would once have been commonplace. In those supposedly 'dark ages'.

In London the planes fly over every two minutes – so even if you listen to a robin singing there is always the sound of a plane overhead in the background. Here the planes go over maybe once every 15 minutes. And there's a road somewhere – but I can't hear it. Instead, there is just the extraordinary song of the birds. On the wall in a little welcome room is a list of the birds that have been seen in the last few days. Later in the day I take a photo so I know that I'm hearing the harmony from this chorus:

1. Yellowhammers
2. Cetti's warblers
3. Whitethroats
4. Swallows
5. Swifts
6. Great tits
7. Blue tits
8. Cuckoos
9. Spotted flycatchers
10. Crows
11. Magpies
12. Jays
13. Linnets
14. Red kites
15. Tree creepers
16. Green woodpeckers
17. Peregrine
18. Hobby (a kind of falcon)
19. Robins
20. Blackbirds
21. Buzzards
22. Kestrels
23. And even a black stork

I imagine it would be the same if you had only ever heard two instruments and for the first time in your life you heard an entire orchestra. And we don't even have to listen: we can hear. It's the experience of being enveloped in a living soundscape that somehow seeps beauty into the human soul.

You know how it is when you sit by the sea and just watch the waves? Even if everything in your life is all wrong, the sound of the waves fills your ears and everything feels right. Here it is again, in a form that I've not experienced before. And I'm not even having to strive to feel peaceful – the sound blanket is real. Eventually a thought arrives: 'This is how it has always been. This is what has existed for decades.' The dark ages were not dark. They were filled with light and sunshine and birdsong, without planes and without cars. The words of the classic song from *Westside Story* come into my head, 'Peace and quiet and open air ... time together with time to spare' Not 'somewhere' or 'someday' as in the song, but right in the present moment and available to us all.

The four of us have dissolved into being. We have ceased the endless mental chatter – the problems of London life that we were all discussing with tears of frustration only hours earlier. We have been absorbed and become the part of nature that we are. All identity with personality, or what the spiritual teachers call the 'self', is gone and we didn't do anything. We are not brought to this heightened experience through any merit. In *The Power of Now*, Tolle speaks of coming to the present. Thich Nhat Hanh teaches the same. Advaita teaches this state. Monotheist religions teach the same, calling it 'the Practice of the Presence of God'. Whether you are believer or atheist, 'being fully here' is being pointed to.

And the birds don't just offer you peace that, for you as the hearer, is effortless. They proclaim peace joyfully.

*

When we finally get up and go for a walk around, we try to put this remarkable experience into words.

'What does being here make you feel?' Zoë asks us.

'It makes me feel tearful,' says Claire.

'Why?' I wonder if it's because we have lost all this. But no.

'Because I'm happy for it.'

'What's "it"?'

'The whole. All of it. What about you, Isabel? Does it make you want to cry?'

I check my heart. 'No – I don't feel tearful. I feel relieved. I feel profound relief. That this exists.'

'Zoë?'

'I feel so calm. It's as if thoughts come in a different way because there is so much that you are sensing and hearing.'

Yes, that's what I'd experienced. An impact on our thinking.

'Emily?'

'I feel reassured and present. And happy. Because when you're here you don't want to be anywhere but here.'

We all agree. So we don't speak any more. We walk and we feed our senses with the giant oaks and the ash and the endless trees that I don't know the names of. We see deer running in and out of clearings, the sudden appearance of two or three young bulls scratching themselves on the trees. The instructions are to keep clear of any large animals we see, but they seem unconcerned by our passing. I love it that they can go wherever they want, wade through a river, have sex, feed their babies with their own milk or take a nap if the mood takes them. They are free.

Every second bush seems to be a hawthorn bush: the blossom that Proust found almost too beautiful for him to look at. He'd have trouble on this walk. It's everywhere.

'You can eat the hawthorn blossoms and the young leaves,' says Claire. 'It's supposed to be good for your heart.'

We walk further in silence.

'I've stopped flushing the toilet,' says Zoë randomly. 'Do you think I have OCA?'

'OCA?' asks Emily.

'Obsessive Climate Anxiety.' Zoë is smiling.

'I don't think so,' I tell her. 'My Grandmother used to say, "If it's yellow let it mellow: if it's brown flush it down." And I'm pretty sure she didn't have climate anxiety. Anyway, my water bills are absurd so it's probably good for your bank account as well as the planet.' Ever the pragmatist.

'Fair enough.'

We walk further in silence, the warm sun on our faces and only the birdsong in our ears.

'Cuckoo,' says Emily.

'Yes, I heard it too,' says Clare. At least we can all recognize that one.

The air feels so good and sweet. We stroll along enjoying every second.

'What's the name of this tree?' asks Emily.

No one can answer.

'But you know – this naming of things …' I reply with the perfect justification for our ignorance. 'Whilst on the one hand it's good to know the names of trees and be able to identify them – and if you also know more about them because of the identification, that's good – but many spiritual teachers also teach that our tendency to name, to label, to identify and tick off, can prevent us from really being with the tree.'

'As we might do with humans?'

'Yes, I guess so. We could say, "Oh, yes – you're Jane Smith. You're a musician," and feel we know who someone is just because we can give them a name. Without stopping to look into Jane Smith's eyes and be with Jane Smith long enough to really feel what it's like to be her and experience who we are in relation to her.'

We walk on a bit further, drinking in the beings. Not naming anything. I ponder aloud, 'The implications of Knepp are so vast. No wonder farmers have strong views. Maybe they don't have to get up at 5am to feed the animals after all.'

'Still farming – but not intensively. I like that. That's a way forward for my conversation with my farmer friends,' says Clare.

We walk through wooded areas and more open areas. We see dug-up banks where the pigs have been playing. We watch the deer watching us before bounding away into the trees.

'Those are fallow deer,' says Clare.

'So beautiful,' says Emily.

That morning we'd met in my local cafe before we set out and had been chatting away like most stressed Londoners. But now we have slowed. Right. Down.

We reach a tree that someone has kindly built a viewing platform around.

There's a notice:

Conservation.

Rewilding is a way of returning former arable land to a state of natural productivity with obvious benefits to wildlife, the soil, local water supplies and, not least, human enjoyment.

The land at Knepp, with its heavy clay and small fields, was never particularly suited to intensive agriculture. It does, however, present ideal conditions for a more extensive system of farming. The method we are practising here is more like ranching, which gives us a significantly lower carbon footprint.

> *A key aspect of the project is carbon sequestration. Permanent pasture woodland and scrub play a vital role in capturing and storing carbon from the atmosphere.*
>
> *The restoration of natural water systems is also important. Bringing back wetlands and floodplains creates another habitat for wildlife. It also helps with water purification and alleviating floods downstream.'*

Amazing – so just by changing the way that they are farming and putting nature in charge, they are also purifying the water and saving the planet.

It's a relief to learn how much can be achieved by trusting nature and doing nothing.

We walk on, barely speaking, until over an hour later we get back to our tent to do nothing there. Really we all need more days like this.

In the evening is the treat that we had booked for: a chance to hear nightingales. These migratory songbirds were once prolific in the UK but populations have declined almost to the point of extinction. The many students that come to research what happens at Knepp noticed that the nightingale populations here are actually going up. So Knepp now has one of the densest populations of nightingales in the UK, with 18 breeding pairs this year.[20] They are astounding little birds – brown, nothing much to look at and just a little larger than a sparrow – but they fly all the way from the UK to sub-Saharan Africa every year. Over 3,300 miles each way. The males arrive first and start singing immediately once they've found a nesting site. The females arrive between ten days and two weeks later. So the male has to sing, for all his little heart is worth, through many solitary nights, to attract a female. Once he's found a mate, he still sings for a while during the day to celebrate (or to protect his territory and let the other males know that this bird is taken). Then, once the chicks are born, he has to stop singing to help with the food gathering. Before setting off on that 3,000-mile flight again.

The evening begins with a delicious candle-lit meal in Knepp's picturesque barn. One of my friends isn't vegan but as I was paying I had ordered four vegan meals. Ha ha. So you can imagine my delight when I find out they had a vegan chef and all the food that night would have been vegan whatever we had ordered. They even have vegan desserts. I'm already in a good mood so this was a welcome bonus. 'Even the red wine is vegan,' said the woman serving proudly. 'Fine – I'll have another glass then.' It would be rude not to.

Then we set off in an old truck to find the nightingales. We have a fair night and so our chances are good. An old friend once told me that a nightingale is very hard to find but our guide was confident. We are driven in one direction but they are so far in the distance they're barely audible. So then we get back in the truck and are driven across the estate to the place where they are singing in their best voices. This serenading only happens for a brief period at the end of May and the beginning of June so you have to know exactly when and where to listen. If you just walk into the countryside not knowing where you are going you're unlikely to hear a nightingale. Here the guides pay attention to nature recovering so they know what is where. The most wonderful part is that Knepp has done nothing to attract the birds. They didn't attempt to create the right kind of undergrowth or grow the right trees. I'll remind you again – they just let the land take charge and the environment for numerous types of birds has created itself.

We get out of our truck close to where the singing is happening. The moon is out so we can see clearly with no torches. You would never see the little male. He's just a tiny speck of brown somewhere in the scrub. But you can certainly hear him. We stand and listen to this beautiful song which is so loud that even a passing female hard of hearing couldn't miss the call. I'm not going to attempt to describe the nightingale's song to you. It's too varied, and I always find the attempts of ornithologists to describe birdsong are absurd. In the case of the nightingale they resort to 'hu-eet', 'jug jug jug jug' and 'pins pins pins' and frankly this doesn't do the bird justice. And, friends, we have the technology. Just open Ecosia and type in 'best nightingale song'. There are two recordings that people have made – each of three hours of singing with no loop on either recording. Then just imagine being outside in nature, with no other sounds except this little voice singing through the night.

For once I hope you're not reading this book far from a computer. Really, no words of mine could do this little bird justice. Please just find these three-hour recordings and enjoy them. At night for full effect. But not with the window open. We don't want to disappoint a passing female.

While you are listening I can let you know that nightingales also sing during the day but as all the other birds sing during the day too you can't hear the nightingale's song as distinctly. So are you listening? Can you begin to define the message to the human heart of this song? My flatmate has just coming running up the stairs as I write this saying, 'What is that amazing birdsong?'

I find myself thinking that they should play this in residential homes for older people. They should play this in hospitals where people are struggling with mental health issues. But then I catch myself. It's not the experience of sitting listening to the recorded version that humans have the right to – but the experience that my friends and I have as we stand here tonight.

A short distance away, another male is competing for the attention of any females flying by in the night. It almost sounds like a call and response as their varied song rings out in the sweet, hawthorn-scented air. But they are not talking to each other, they are each singing their own energetic mating song.

'Can I move closer?' I ask our guide.

'Move as close as you like,' she says. 'They're not bothered by us.'

I move forward until I'm standing so close that the song is perfectly clear and I can hear every tiny note. Except apparently I can't. Some of the notes the birds sing are too high to be heard by the human ear. They are for the birds' ears only.

He sings on and on. He'll sing all night like this, every night, for as long as it takes. We stand listening. Each of us is loath to move on from this magical present moment. We each listen with our own thoughts or, more interestingly perhaps, each without our own thoughts. In this moment we are helped to just be here because even listening isn't required. Only being here.

Eventually we turn and are driven back to where our tent is pitched so we can sleep. As we lie on our sleeping mats, in the far distance you can still hear him. Bless his heart. I hope they attract a female, each and every one. And that, thanks to Knepp, their numbers recover, farmers and landowners follow suit and that somehow this beautiful sound will once again be familiar. I can dream at least.

And I do dream – of beautiful birds in a restored natural landscape. Until the following morning when the most enthusiastic and prolific dawn chorus we've ever heard, with every bird at Knepp joining in, joyfully announces the arrival of another glorious day.

POSTSCRIPT
And Where From Here?

Once long ago, but in this lifetime, I had the privilege of interviewing His Holiness the XIV Dalai Lama of Tibet. For another book – *For Tibet, With Love* – I had been involved in exploring the age-old question, 'What can one person do to make a difference?' During the project I had used for guidance the famous 'Serenity Prayer', 'Grant that we may have the serenity to accept what we cannot change, the courage to change what we can and the wisdom to know the difference.' I had given long consideration to serenity and courage – what I lacked totally was any wisdom.

I was overawed to meet the Dalai Lama and mumbled my long question – which bewildered both His Holiness and the translator.

My question sounded something like: 'There is a famous prayer, your Holiness, in which, er, we consider, er, how we might learn to cope with things that we can't do anything about and, er, gain some serenity in those circumstances. Which is hard of course. The way the world is. And, er, on the other hand the prayer says that, er, we need to have courage to change things we can change. And there is an assumption, in the prayer, er, your Holiness, that if we have courage, we can change stuff, er, I mean some things. And, er, your Holiness, the third part is about, er, wisdom, to know the difference, er, between these two. And, er, as I know that you are, er, that Tibetan Buddhists believe you to be, er, an incarnation of wisdom … so I was wondering, er … how we can learn to discern the difference between what we can make a difference to and what we probably can't change, er, your Holiness, because it's outside what we think we can influence.'

The Dalai Lama looked bemused. He turned to his translator. They conferred in Tibetan. They smiled. The Dalai Lama turned back to me and he said, 'Experiment.' That's the difference between a not-so-lucid mind and a lucid one.

And with that one word he gave me – and all of us – a game to play for the rest of our lives.

So, please – I invite you to experiment with everything in this book. Some of the ideas here will work for you and some won't. In many areas you will have much better ideas than I have. Please write if you do and, if possible, I'll include your suggestions in any future editions of this book.

With all the ideas you have for things outside your own life and your own home, experiment there too. The world is our playground. Sometimes if you see a solution, others will instantly say, 'That's a good idea,' and change something. Even climate sceptics are open to environmental solutions that also save money.

If you don't have an environmental charity that you love, then adopting a favourite one is a good excuse to do all kinds of crazy fun stuff to raise money for them. We rarely feel powerless when baking, walking over fire, enduring a triathlon or doing whatever we most enjoy doing to raise money. Campaigning with thousands of others to change the political conversation is also a powerful antidote to feeling that one person is not enough.

If you spot an environmental problem and can see a solution, write a letter and experiment with whatever produces results. Perseverance is another powerful tool. Perseverance, joy, perseverance, joy … keep going.

When you remain focused on solutions, you don't need to become discouraged. If you try something and it doesn't work, simply experiment with a different approach. It's not knowing what to do that leads to people feeling powerless. But, as I hope you've seen, there are many actions we can take and not take – every single day. And if we can enjoy our choices, we can be effective and not get burned out. Of course, when we experiment we'll often fail. And that's OK too. Drink tea. Start over.

I'm looking forward to continuing this journey for the rest of my life, to supporting others that are travelling the same route and to applauding others as they find solutions too. I've listed a good number of these inspiring beings in the Appendices. Please follow them. Follow me too.

Think Globally – Act Joyfully.

Thank you for travelling with me.

x Isabel

APPENDIX 1
Further Reading

Ten more books you might like to read on this subject. In no particular order.

1. *Feral* by George Monbiot. George understands everything and has a brain like a planet. It's a torment to him as he sees solutions to problems so easily. Above all, this book gives you a feeling and a reminder of what our relationship with nature could be.
2. *This Changes Everything* by Naomi Klein. If you want to increase your understanding of how climate change has been caused, the influence and power of the fossil fuel industry and how we got into this mess, this one is vital. I read it twice, slowly and took notes. Sometimes it's a pleasure to read Klein no matter how difficult her subject material, just because she writes so well. But it's certainly a read that requires concentration.
3. *Wilding* by Isabella Tree. The full story of Knepp and the rewilding of their land. Truly inspirational and an easy read. If you know any farmers, please buy them this book.

4. *This is Not a Drill: An Extinction Rebellion Handbook.* All about Extinction Rebellion: what they are doing, how and why they are doing it – and inspiration to join in.
5. *Doughnut Economics: Seven Ways to Think like a 21st-century Economist* by Kate Raworth. I have to confess I didn't read this one. I downloaded it and listened to it. I'm not an economist but this isn't hard to understand. If you want to get your head around what is wrong with our current rampant model of capitalist profit and how we can fix it, this will be a good one for you.
6. *There Is No Planet B: A Handbook for the Make or Break Years* by Mike Berners-Lee. This book has a great format: the entire book is in a question-and-answer structure so it makes it easy to look things up. It's full of facts, figures and statistics. If you have a mathematical brain and fully comprehend the significance of statistics, then this is a brilliant book for you. If, furthermore, you can retain the figures then this book will arm you with enough information to convince anyone of the severity of the crisis. I listened to Mike reading his own book and he comes across very well – with a comprehensive understanding but also down-to-earth and friendly.
7. *Climate Justice: A Man-Made Problem with a Feminist Solution* by Mary Robinson. Robinson is a UN Special Envoy for Climate Change and a former UN High Commissioner for Human Rights. With stories from around the world from people who are already living with the consequences of climate change, she demonstrates why Safia is right when she talks about our responsibility as consumers.

8. *How to Save the World for Free* by Natalie Fee. A highly accessible and readable handbook from the founder of City to Sea, the charity that campaigns to get plastic out of our oceans. In short sections and even has pictures – beautifully presented and full of good ideas and practical suggestions.
9. *No One Is Too Small to Make a Difference* by Greta Thunberg. A tiny book of 68 pages which gathers together all Greta's speeches from climate rallies across Europe to audiences at the UN, the World Economics Forum and the British Parliament. She's brilliant. She says 'We must stop the emissions' every time she speaks. Then she adds, 'Is my microphone working? I wonder if you can hear me.'
10. *Client Earth* by James Thornton and Martin Goodman. A book that makes reading about Environmental Law exciting and compelling. If you know a student that is thinking of studying law – please give them this book. Published in 2017 and (sadly) even more important today than it was then.

APPENDIX 2
Further Cooking: Fifteen Favourite Plant-Based Cookbooks for Flavoursome Inspiration

Fifteen because it was impossible to stop at ten – they are all so inspiring. Some here are recommended by me and some by friends.

1. *Vegan Recipes from the Middle East* by Parvin Razavi. Includes dishes from Iran, Armenia, Syria, Lebanon, Jordan, Egypt, Morocco and Turkey. Just so much variety and deliciousness. If people ask, 'What do vegans eat?' just show them this book.
2. *The First Mess Cookbook: Vibrant Plant-Based Recipes to Eat Well Through the Seasons* by Laura Wright. A beautiful plant-based cookbook with the added

advantage of the seasonal aspect of what is available when.

3. *Vegan Street Food: Foodie Travels from India to Indonesia* by Jackie Kearney. Winner of the PETA award for Vegan Cookbook of the Year. Jackie and her family ate their way around Asia, sampling street food and jotting down menu ideas on the back of napkins.
4. *15 Minute Vegan: Fast, Modern Vegan Cooking* by Katy Beskow. The title does it for me. I'm about to order this from my local library and try out everything in it. Katy has also compiled a book called *15 Minute Vegan Comfort Food*. I wish she lived in my house.
5. *Afro-Vegan: Farm-Fresh African, Caribbean, and Southern Flavors Remixed* by Bryant Terry. If you like Caribbean food and want to be vegan then this is the book for you. Contains recommendations of what to listen to while cooking. I just love it.
6. *Bosh! How to Live Vegan: Save the Planet and Feel Amazing* by Henry Firth and Ian Theasby. These pioneers of simple, easy and delicious plant-based cooking help make a sustainable and ethical lifestyle easier. Clear and well set out. (Recommended by Jo McDonald and Sarah Rose.)
7. *Fresh India: 130 Quick, Easy and Delicious Vegetarian Recipes for Every Day* by Meera Sodha. I know this is vegetarian and not vegan but I'm including it here as it has long been a favourite in this house. It's just so beautiful, both visually and in taste if you love Indian flavours. (Recommended by Emily Lucienne.)
8. *Spring and Summer: Cooking with a Veg Box* and *Autumn and Winter Veg* from Riverford. Two books that they supply to show you what to do with the

seasonal fruit and veg that they are delivering. (Recommended by me.)
9. *Eat Well Live Well with The Sensitive Foodie: An Accessible, Practical Guide to Eating a Whole-food Plant-based Diet with Over 100 Delicious Recipes* by Karen Lee. A non-vegan friend recommended this one and adds that she appreciates not only the recipes but also the supportive science behind the choices. (Recommended by Phillipa King.)
10. *Veganomicon: The Ultimate Vegan Cookbook* by Isa Moscowitz and Terry Romero. Published in 2007 and still the favourite of many plant-based foodie friends. (Recommended by Mandy Ollis and Audrey Ljn.)
11. *The How Not to Die Cookbook: Over 100 Recipes to Help Prevent and Reverse Disease* by Dr Michael Greger and Gene Stone. Based on the bestselling *How Not to Die*, this links specific foods with specific diseases and tells you how to sort your health out by eating amazing food. A great gift for anyone with chronic health challenges. (Recommended by Susie Beaumont.)
12. *Vegan One Pound Meals* by Miguel Barclay. An excellent gift for anyone who complains that a plant-based diet is expensive. This is not the most health-conscious of the options here – and there has been some criticism that the recipes are heavy on the syrup – but borrow it from the library if you're skint and find out which of the £1 meals you love. Getting a book from a library supports the author and the library. (Recommended by Lee Walls.)
13. *Vegan for Good : Deliciously Simple Plant-Based Recipes for Every Day* by Rita Serano. Some vegan cookbooks, I have discovered, are not actually written by vegans

but Rita Serano is the real deal. This book is beautiful, not too expensive and each of the recipes has seven ingredients or less, so it's not too complex either. (Recommended by Stuart Lemon.)

14. *Jackfruit and Blue Ginger : Asian Favourites, Made Vegan* by Sasha Gill. This is one for those of you who really love cooking. When Sasha became vegan she didn't want to miss out on all the delicious flavours of her heritage so she set about 'veganizing' them. I have just opened my copy at random: Pandan and coconut dumplings. Yes please. Great photos in the book too. (Recommended by Jane Graham-Maw.)

15. *The Homemade Vegan Pantry: The Art of Making Your Own Staples* by Miyoko Schinner. A whole book of making vegan staples like milk, butter, salad dressings, pesto, mayo, lemon curd, bread, pancakes, cheese sauce, cheese, cream, sausages ... basically all of those things that are usually highly processed. This book teaches you how to do it yourself with all natural ingredients. (Recommended by Helen Estevao.)

APPENDIX 3
Ten Environmental Charities

If you want to give support to a charity that is working to support the planet, here are some – both small and large – that you may like to choose to raise money for or become a regular donor to.

If you don't already give, choose one and give at least something. Small amounts from large numbers of people make a difference. Even if it's just the price of a coffee once a month – £2.50, say. And if you can't, that's fine – leave it to others who can or wait until a time in your life when you're more financially abundant.

Just learning about the work of these charities, non-profits and other organizations can restore your faith in humanity.

1. The Environmental Law Foundation. A charity that guides, helps and empowers activists and environmental groups.
2. The Devon Wildlife Trust. 'Our vision is a Devon that is rich in all wildlife where people can enjoy and care

about the natural world and take steps to safeguard it for the future.'
3. Greenpeace International. Still one of the most internationally admired ecological charities. It is criticized by some for not being radical enough – but when was the last time the critics took a ship up to the Artic to put their own bodies between the fossil fuel industry and the ice?
4. Derbyshire Wildlife Trust. Supporting, monitoring and developing campaigns and policy for wildlife for over 55 years.
5. The Shark Trust. We need sharks to keep our oceans healthy but human activity, false Asian ideas about virility and shark fins, misrepresentation by Hollywood and human greed is pushing these species to the brink of extinction. The Shark Trust – a tiny charity – does what it can to fight back.
6. Extinction Rebellion. Not a charity, more a mass movement of individuals who have given up campaigning by more conventional means. If you go to their website you can contribute either to their actions or to the legal costs of people who have chosen to be arrested in conscious acts of civil disobedience.
7. Global Canopy. Works to protect the world's ancient and endangered forests. They work with the forest industry's customers and their suppliers to develop business solutions that protect the forests.
8. Sea Shepherd. An international non-profit marine wildlife conservation organization that uses direct action to confront illegal action on the seas.

9. Bug Life. The only organization in Europe devoted to the conservation of all invertebrates. Because we have to love the insects.
10. Trees for Life. Rewilding the Scottish Highlands.

APPENDIX 4

Twelve Instagram Accounts of People or Organizations Mentioned in this Book

1. Trees for Life @treesforlifeUK
2. Triodos Bank UK @Triodosuk
3. Riverford @riverford
4. Good Energy @goodenergyuk
5. Lammas @lammasearthcentre
6. Fridays for Future @fridaysforfuture; Extinction Rebellion @extinctionrebellion
7. People Tree @peopletreeuk; Po-Zu shoes @Po_zu
8. The Green Party of England and Wales: @thegreenparty
9. Isabella Tree @isabella.tree; Knepp Wildland Safaris @Kneppsafaris
10. Canopy @canopyorguk The National Association of Tree Campaigners
11. Ethical Consumer Magazine @ethical_consumer_magazine
12. Heat Geek @heat_geek

APPENDIX 5
Ten More Environmental Instagram Accounts to Follow (+ mine)

1. The Soil Association @soilassociation
2. Sarah Aubrey @electrify_this
3. Stop Ecocide @ecocidelaw
4. Ecosia @ecosia
5. Ben Goldsmith @bengoldsmith
6. Two Minute Beach Clean @2minutebeachclean
7. Flight Free 2020 @flightfree2020
8. Environmental Education @nowhereandeverywhere_
9. John T McFarlane @johntmacfarlane
10. Save Windermere @mattstaniek
11. Isabel Losada @isabeljmlosada

Acknowledgements

I would like to thank everyone who is mentioned in this book and/or who has played a part.

Especial thanks are due to all the staff and volunteers at Trees for Life in Scotland, Guy Singh-Watson and everyone at Riverford and all those that give their time to Moor Trees in Devon. Thank you to Sarah Ives for her enthusiasm for non-car use and to Rosalind Redhead, who, although not in the book, campaigns constantly for cities free from private car use. And thank you to Simon Birkett, who, although likewise not mentioned here, campaigns with equal vigour for cleaner air.

It may seem unlikely but I feel I should thank Marie Kondo. Although I'm not sure Ms Kondo intended to be an environmentalist, by teaching so many of us to live with fewer objects and to truly value what we have, she has become one. I would also like to thank everyone who attended one of my weekend workshops and shares my enthusiasm for living with less.

It's a privilege to be able to interview the CEO of my energy provider – and what a joy to find a warm, funny and committed environmentalist running the show. Meeting Juliet Davenport was a complete delight.

Thank you to the many friends and strangers I have rung up for advice at various points in this project: Patrick Uden, Jenny Hanafin – the pregnant climate scientist (who now has a little boy, Cillian) – the RHS, the RSPB and Natalie Fee at City to Sea.

Those that I met at Extinction Rebellion are too many for me to list or even to remember – apart from Jake and Pedro, who I visualize both conducting a motley crew of non-professional drummers while a police officer tries, rather half-heartedly, to arrest them. On behalf of the planet (if I may? Just speaking as one human?) I'd like to thank every single person who showed up and will continue to show up, to prove how much we care about our planet. Thank you most especially to the 'arrestables' whose courage and determination really has made a difference and will continue to do so. You are all my heroes.

Thank you to Caroline Lucas, Greta Thunberg and Anna Taylor. Just thank you.

I am grateful to the founders and staff at Triodos for giving me a bank I am proud to bank with.

Locally, I am lucky to have a wonderful Green Party, and an active local Extinction Rebellion Group with whom I play wonderful games (who can't love putting a yellow ticket on an SUV, telling them that driving such vehicles in cities is an environmental crime?) or guerrilla gardening (taking away litter in the dead of night and planting flowers instead). These wonderful humans collected 10,000 signatures demanding that their local council take action for the climate. And Wandsworth Borough Council is listening and taking action.

Thank you to everyone who is kind to anyone who bangs on their door to canvass for a political party they believe in

ACKNOWLEDGEMENTS

– especially if they are representing a party that you don't vote for.

Thank you so much to Tao, Hoppi and all those living in or around Lammas that were so helpful, and all those in my group the week of my visit. I hope all your plans to escape the rat race and live off grid with nature are going well. This book is also dedicated to all of you.

Where our clothes are concerned, many have been campaigning for decades against the excesses of the fashion industry. Thank you so much to Safia Minney for being one of those leaders on the front line and for her inspiration.

My visit with Clare, Zoë and my daughter to the Knepp Estate in West Sussex, UK was the high point of this project for me because we saw what is possible if we can just leave nature alone and stop trying to control her. Thank you so much to Isabella Tree and Charlie Burrell for their vision.

I would like thank my agent, Toby Mundy, for his humour, and the staff at Watkins for being so dynamic. Publisher Fiona Robertson, copyeditor Sophie Elletson and the astounding head of PR and marketing Vicky Hartley: all worked so hard and published with fantastic speed. Thank you also to all the other staff at Watkins who I don't meet personally but who have played a part. I'm most grateful to Etan Ilfeld for making everything possible.

A particular word about the cover, designed by the brilliant Jonny Hannah who has captured all the joy and energy for the cover I was hoping for. Jonny has designed covers for my books (except *The Battersea Park Road to Paradise*) since Will Webb matched us up at Bloomsbury many years ago. I am so grateful to Jonny – I love his crazy and wonderful creativity. I hope you like the cover of this edition as much as I do.

On the subject of social media, a lot is said about the evils of it. But when you are an author, which is mostly an isolating job, contact with others across the world is very much appreciated. So if you follow me on FB or Instagram – I'm grateful for each and every supportive click and encouraging comment. The vegan and waste-free communities on Instagram have taught me so much and I learn every day from the accounts of those that I admire.

Closer to home I'd like to thank my daughter for her long-standing toleration of her mother's activities. The news that I'm out narrowly avoiding arrest is nothing compared to the news of some of the activities I was exploring in the last book – which she read with one eye closed. Emily's patient tolerance and love for her mother is like the sun – huge. Other key supporters of this book are Jane Stephenson, Gala Riani, Stuart Lemon, Liz Pearce and Kate Latham, all who helped me by reading and providing valuable guidance. Thank you so much you wonderful people.

Book blurbs are considered important in publishing so to give one is a huge kindness. I know you aren't supposed to put people on pedestals but frankly some people merit one. Although each of these great souls has so much humility that they would never agree to stand on one. They are some of the best teachers I know of what it is to be human. I'd like to thank Dr Rowan Williams, George Monbiot, Isabella Tree and Sir Mark Rylance for their commendations. And, well, for existing.

Lastly – if you are holding this book or listening to this book, I thank every one of you for your determination to be part of the solution. Thank you so much. Xxx i.

Endnotes

1   Matthew Taylor, *The Guardian* (2017) https://www.theguardian.com/environment/2017/dec/26/180bn-investment-in-plastic-factories-feeds-global-packaging-binge

2   After Brexit this is no longer true. The rules were changed after Brexit. There is hope that in the future subsidies will be payable for growing scrub and creating other biodiverse environments.

3   1.35 million a year according the World Health Organization.

4   The non-profit environmental group City to Sea reports that 980 tonnes of mini plastic 'travel-sized' bottles are dumped abroad by British holiday makers every year.

5   The UK's last coal-fired power station at Radcliffe-on-Soar, near Nottingham was closed in September 2024 between the first and second editions of this book.

6   I enquired with my local council who told me that I would be 'very unlikely' to get permission. Even if I did have either the money or the desire to stick something on the front of my Victorian terraced house.

7   https://phys.org/news/2018-06-years-earth-due-greenhouse-gas.html

8   *www.noaa.gov/news/carbon-dioxide-levels-breach-another-threshold-at-mauna-loa*

9   https://www.ipcc.ch/about/structure/

10   https://www.youtube.com/watch?v=PTksae5SNQk

11   From 2024 Brighton Pavilion has been represented by Siân Berry – also of the Green Party.

12   At the time of the second edition of this book going to print, Juliet now works as an energy transition advisor, including for the 'Clean Power 2030 mission'.

13   In 2023 UK electricity generation was 43 per cent zero carbon.

14   Good Energy will now install solar panels and heat pumps providing you live in the right part of the country. You can go to their website and put in your postcode to see if they cover you.

15   You can certainly get a second generation smart meter which gives live data on how you're using your energy via an app on your computer.

16   Green Energy is now known as 100 Green.

17   Research from the Global Footprint Network.

18   At the time of writing, locally to my home in London, Waitrose is the only place I've found where I can buy English red wine, and it's not organic.

19   There are now 19 locations for 'Library of Things' in London alone and petitions to start them all over the country. Find them at libraryofthings.co.uk.

20   In 2024 there were 60 breeding pairs.

The manufacturer's authorised representative in the EU for product safety is EU Responsible Person (for authorities only) – eucomply OÜ – Pärnu mnt 139b-14, 11317 Tallinn, Estonia, hello@eucompliancepartner.com, www.eucompliancepartner.com